**The Technol**

This book is for:

Ruth, Sally, Jasmine and Jack
Rosie, Ruth and Rebecca
for all their support and understanding.

# The Technology of Building Defects

John Hinks
Geoff Cook

E & FN SPON
An Imprint of Thomson Professional
London · Weinheim · New York · Tokyo · Melbourne · Madras

**Published by E & FN Spon, an imprint of**
**Thomson Science & Professional, 2–6 Boundary Row, London**
**SE1 8HN, UK**

Thomson Science & Professional, 2–6 Boundary Row, London
SE1 8HN, UK

Thomson Science & Professional, GmbH, Pappelallee 3,
69469 Weinheim, Germany

Thomson Science & Professional, 115 Fifth Avenue, New York,
NY 10003, USA

Thomson Science & Professional, ITP-Japan, Kyowa Building, 3F,
2-2-1 Hirakawacho, Chiyoda-ku, Tokyo 102, Japan

Thomson Science & Professional, 102 Dodds Street, South Melbourne,
Victoria 3205, Australia

Thomson Science & Professional, R. Seshadri, 32 Second Main Road,
CIT East Madras 600 035, India

Distributed in the USA and Canada by Van Nostrand Reinhold,
115 Fifth Avenue, New York, NY 10003, USA

First edition 1997



Typeset in 10/12 Cheltenham by Photoprint, Torquay, Devon
Line illustrations redrawn from the author's originals by Passmore
Technical, Falmouth, Cornwall.
Printed in Great Britain by Alden Press, Osney Mead, Oxford

ISBN 0 419 19780 X



A catalogue record for this book is available from the British Library

♾ Printed on acid-free paper, manufactured in accordance with
ANSI/NISO Z39.48-1992 (Permanence of Paper).

# Contents

## PART TWO COMPONENTS



## PART THREE ELEMENTS

# Acknowledgements

We are indebted to the following for permission to use copyright material: Dr W.A. Allen of Bickerdike Allen Partners for the thermal pumping concept and diagram; Jack Hinks for the photograph of tree damage; the *Kentish Gazette* for the photograph of wind damage; Addison Wesley Longman Ltd for permission to reprint several items from Cook and Hinks, *Appraising Building Defects;* and Keith Bright, Diane McGlynn, Mr S.D. McGlynn and the University of Reading for permission to reprint several of their photographs.

# Introduction

This book provides a substantial stand-alone review of the technology of building defects. The text is extensively illustrated photographically and diagrammatically.

General educational objectives are provided which offer the reader the opportunity of self-assessment. Each section is a self-contained review offering accessibility to the reader across a range of technical topics concerned with building defects. The book can be used for direct lecture material, seminar/tutorial information, assignment work and revision notes. In this way it provides a one-stop source for student-orientated knowledge which is all too often hidden in a mass of different information. The book provides the reader with sufficient detail of the technology of building defects to enable them to be placed in an overall context.

Further reading to expand the content of the topic areas is listed.

PART 1

# Materials

CHAPTER 1

# Defects with materials

# 1.1 Material properties generally

**Learning objectives**

- You will be made aware of a range of physical properties of common building materials.
- You should recognize that these properties are essential features in the understanding of deficiency in buildings.

## Material properties generally

The properties of materials shown in Table 1.1 are for general guidance only. There are specific properties for particular materials and further information can be obtained from the '**Further reading**' list.

The 'melting points' given in Table 1.1 for GRP, unplasticized PVC and plasticized PVC are, strictly, inappropriate. This is also true for glass, which does not have a well-defined melting point. The temperature shown is where the material starts to move as a fluid.

Although reversible shrinkage has been tabulated here, several materials also exhibit irreversible shrinkage.

Where there is a dash in the table this generally indicates that within the general description of the material there are a variety of different forms available. This in turn means that there are a wide range of published values for this material. The properties of timber are described in section 4.1 of this book.

---

## ■ Discussion topics

- Describe the influence of physical factors on the deterioration of three building materials.
- Discuss the use of the term 'lack of strength' to describe the deterioration of building materials.
- Explain the anisotropic nature of timber.
- Compare the influence of chemical properties on the deterioration of ferrous metal and aluminium.

*The Technology of Building Defects.* Dr John Hinks and Dr Geoff Cook.
Published in 1997 by E & FN Spon, 2–6 Boundary Row, London SE1 6HN, UK. ISBN 0 419 19770 2

**Table 1.1** General properties of common building materials

| Material | Coefficient of linear expansion ($10^{-6}$ per °C) | Unrestrained movement for 50°C change (mm/m) | Moisture movement, reversible (%) | Density (kg/m$^3$) | Failure stress (N/mm$^2$) | | | Melting point (°C) | Thermal conductivity (W/m k) |
|---|---|---|---|---|---|---|---|---|---|
| | | | | | Compression | Tension | Bending | | |
| Bricks and tiles (fired clay) | 5–6 | 0.27 | 0.02 | 1 975 | 20 | – | – | – | 1.15 |
| Limestone | 6–9 | 0.40 | 0.08 | 1 950 | 16.5 | – | – | – | – |
| Glass | 7–8 | 0.35 | None | 2 560 | 15 | – | – | 1500 | 1.05 |
| Marble | 8 | 0.40 | Negligible | 2 880 | >100 | – | – | – | – |
| Slates | 8 | 0.40 | Negligible | 2 950 | 53 | – | – | – | – |
| Granite | 8–10 | 0.40–0.5 | None | 2 850 | >170 | – | – | – | – |
| Asbestos cement | 9–12 | 0.45–0.6 | Negligible | 1 700 | – | – | 19 | – | 0.58 |
| Concrete and mortars | 9-13 | 0.55 | Negligible | 2 400 | 20 | 2 | 4 | – | 1.60 |
| Mild steel | 11 | 0.55 | None | 7 850 | – | >500 | 165 | 1900 | 50 |
| Sand–lime bricks | 13–15 | 0.70 | 0.025 | 1 975 | 20 | – | – | – | 1.15 |
| Austenitic stainless steel | 17 | 0.85 | – | 7 850 | – | >500 | – | 1050 | 15 |
| Copper | 17 | 0.85 | – | 8 940 | – | 280 | – | 1083 | 400 |
| GRP | 20 | 1.00 | – | 1 602 | – | 138 | – | 93 | – |
| Aluminium | 24 | 1.20 | – | 2 650 | – | 100 | 162 | 660 | 214 |
| Lead | 29 | 1.45 | – | 11 340 | – | 15 | – | 327 | 35 |
| Pure zinc | 31 | 1.55 | – | 7 140 | – | 150 | – | 419 | 113 |
| Unplasticized PVC | 50 | 2.50 | – | 1 394 | – | 55 | – | 56 | 0.16 |
| Plasticized PVC | 70 | 3.50 | – | 1 281 | – | 12 | – | 40 | 0.4 |
| Polycarbonate Polythene | 70 | 3.50 | – | 1 190 | – | 60 | – | 130 | 0.23 |
| (HD) | 143 | 7.00 | – | 945 | – | 29 | – | 94 | 0.50 |
| (LD) | 198 | 9.50 | – | 913 | – | 11 | – | 72 | 0.35 |

- 'Because modern construction makes use of high-quality materials the incidence of defects will inevitably decline'. Discuss.

---

## Further reading

Cook, G.K. and Hinks, A.J. (1992) *Appraising Building Defects: Perspectives on Stability and Hygrothermal Performance*, Longman Scientific & Technical, London.

Curwell, S.R. and March, C.G. (eds) (1986) *Hazardous Building Materials: A Guide to the Selection of Alternatives*, E. & F.N. Spon, London.

Desch, H.E. (1981) *Timber: Its Structure, Properties and Utilisation*, 6th ed (revised by J.M. Dinwoodie), Macmillan Education, London.

Everett A. (1975) *Materials – Mitchells Building Construction*, Longman Scientific & Technical London.

Richardson, B.A. (1991) *Defects and Deterioration in Buildings*, E. & F.N. Spon, London.

Taylor, G.D. (1991) *Construction Materials*, Longman Scientific & Technical, London.

# 1.2 Stone

## Learning objectives

You should be able to:

- compare the influence of moisture and thermal movement on the durability of stones used in construction;
- explain the methodology of chemical deterioration of building stones;
- compare the chemical deterioration of different building stones;
- describe the influence of porosity on the degree of exposure permitted for limestones;
- describe the influence of cleavage planes on the durability of slate, shale and sandstone.

Types of stone can be classified with respect to age, composition, behaviour and location. In general the difference associated with age is the accepted method of describing building stones. There are a large variety of building stones used in construction; these include igneous, sedimentary and metamorphic stones.

## Deterioration of igneous stone

This stone is produced by cooling of fluid from the interior of the earth. This 'magma' can be 500 to 600 million years old or may be more recently formed from volcanoes. This cooling may be in the air (extrusive igneous rocks), or underground (the 'intrusive' or 'plutonic' igneous rocks). These rocks cool more slowly and are generally more crystalline, e.g. granites.

Igneous stone can be hard, durable, impermeable and inert. Some basalts and dolerites have significant moisture movement characteristics causing a moisture movement coefficient of 0.1% to 0.3% in concrete. The deterioration mechanisms associated with igneous rocks are commonly associated with brittle failure or surface discoloration caused by weathering or condensation. They are commonly of a minor nature.

*The Technology of Building Defects.* Dr John Hinks and Dr Geoff Cook.
Published in 1997 by E & FN Spon, 2–6 Boundary Row, London SE1 6HN, UK. ISBN 0 419 19770 2

## Sedimentary stone – general

The compaction of weathered igneous rocks or shell fragments produces sedimentary rocks. In general the compacted shell fragments produce limestones and the weathered igneous rocks produce sandstones.

### Problems with limestones

The limestones, which are very common in the UK, are predominantly composed of calcium carbonate. Other compounds, e.g. magnesium carbonate, may be present and these give particular characteristics to the limestone. Porosity can vary between 1% and 40%. This and the saturation coefficient are coarse measures of durability, which for limestones can be more accurately assessed by quantifying the percentage of pores below 5 microns, where $<30\% =$ durable, $>90\% =$ not durable. A crystallization test can be carried out on limestone samples. They are soaked in a sodium sulphate solution and dried 15 times, and the resultant effects are used to classify the stone on a six-point scale from 'A' to 'F'. These are used to define appropriate exposure zones and can be used to identify inappropriate applications.

**Table 1.2** BRE exposure zones for limestones

| Zone | Description | Lowest suitable class | Permitted exposure |
|---|---|---|---|
| 1 | Paving, steps | A | Any |
| 2 | Copings, parapets | B | Pollution and/or coastal exposure |
| 3 | Quoins, strings | C | Neither of the above<br>Pollution or coastal exposure |
| | Mullions, cills | D | Neither of the above |
| 4 | Plain walling | E/F | Neither of the above nor pollution exposure |

Limestone is chemically more active than sandstone. The quarrying of limestone, which contains a mineral- and salt-rich moisture, may be followed by 'seasoning', where the moisture moves to the surface. Evaporation of the water leaves a crystalline crust which is removed when the stone is 'dressed'. Dressing freshly quarried stone is considered to be easier, although the crystallization will now occur on the finished stone. In general the lower the moisture content of the stone then the greater the frost resistance.

Limestone can deteriorate when exposed to acidic rainwater or any other sulphurous source. Carbon dioxide when dissolved in rainwater has a pH of around 5.6, whereas acid rain has a pH of less than 5. The sulphurous acids can combine with the calcium carbonate to produce calcium sulphate and

**Fig. 1.1.** The possible organic attack of stonework or of other material behind the stonework. Fruiting bodies may develop away from the direct region of attack. (S.D. McGlynn.)

calcium nitrate in the surface region of the stone. Whilst this may be hard and dense it has different physical characteristics from the base stone and the resultant differential moisture and thermal movement, together with the stresses associated with its crystallization, can cause the stone face to break down. In addition calcium sulphate is slightly soluble, causing the gradual erosion of the stone surface under the action of rainwater. The calcium nitrate is hygroscopic and will absorb water from the air. This will drive the deterioration process and accelerate the deterioration of the stone.

There are also problems associated with the bacteria, algae, fungi and lichens commonly present on the stone surface. Some of the bacteria have the ability to convert the sulphurous and nitrous acids from environmental pollution to sulphuric and nitric acids, which can be more damaging to the stone.

Water run-off from limestone can cause pattern staining of façades and may also cause deterioration of sandstone. Additional staining may come from chemicals washed from adjacent materials, e.g. green staining from copper.

## Problems with sandstones

The sandstones are commonly held together with a silica or calcium matrix, in well-defined bedding planes of differing composition. Where quartz is held together with a silica matrix this is termed silicaeous sandstone; with a calcium carbonate matrix, calcareous sandstone; or where calcium carbonate and magnesium carbonate form the matrix, these rocks are termed dolomitic sandstones.

The porosity of these sandstones can vary between 1% and 25% and saturation coefficients between 0.5 and 0.7. Since this variation can occur within similar sandstones the weathering performance and frost resistance of the stone may vary, even on the same building.

These stones may also be damaged because of crystallization of calcium sulphate or calcium nitrate within the stone and below the surface. Atmospheric pollution can cause deterioration of sandstone when the sulphurous gases in the air condense to react with the calcium carbonate, producing soluble calcium sulphate. The attack mechanism is similar to that which affects limestones. Water run-off from limestone can cause contour scaling of a sandstone surface. The water run-off will enrich the sandstone surface and underlying layers with calcium sulphate and calcium nitrate. The calcium sulphate and calcium nitrate enrichment of the surface crust causes differential thermal and moisture movement problems. This can result in the surface becoming detached in a manner which follows the contours of the stone surface.

In external applications delamination can occur where bedding planes have been laid vertically, parallel to the building façade. Where the bedding planes are laid horizontally in highly carved stones, there is a risk of localized deterioration where the weathering of exposed and unrestrained bedding planes can occur.

Dissolved salts from groundwater and those from sea spray can cause disruptive damage to limestones and sandstones. Where sodium chloride or sodium sulphate has crystallized the effects are likely to be widespread across the surface of the stone.

## Problems with metamorphic stone

The modification of rocks or other material by heat and pressure can produce hard, durable and attractive rocks. These are termed metamorphic rocks and may also be composed of older igneous rocks, e.g. granites. This rock classification includes the slates, which are formed from clay, and marble, which is formed from calcium-rich rocks such as limestone and sandstone. Low compaction, as in the case of shales, can lead to a likelihood of moisture movement, which, where cleavage planes are disrupted, can be irreversible.

The pronounced laminations, or bedding planes, within a slate mass allow the stone to be split into thin sections. This is particularly useful for roofing applications, where thicknesses of 3 to 5 mm are common. Whilst there are standard tests for the durability of slates, a good guide is to appraise the performance of the same slates in use. The tests include water absorption, which is usually very low, acid resistance, which is usually very high, and resistance to delamination when exposed to wetting and drying cycles. These tests can be carried out on existing slates although an allowance must be made for the deterioration of performance due to aging of the slate.

Marble is commonly used for its appearance. This can be enhanced by streaks and coloured patterns of other minerals in and around the base calcareous rock and by polishing. The polished surface can lose its sheen because of weathering when used externally, since marble can very occasionally be attacked by sulphurous gases.

## General deterioration factors

Where stones, e.g. limestone, marble and granite, are used in thin sections for external cladding they must be tolerant of a wide temperature range and in some cases rapid temperature variation. In particular there is a need to consider the effects of thermal expansion where the stone is attached to a concrete structure. An average coefficient of thermal expansion of limestone is 4.0 to $8.0 \times 10^{-6}$ per °C, and this compares to an average value of 9.0 to $12.0 \times 10^{-6}$ per °C for gravel aggregate concrete. Differential thermal movement is less marked for sandstone and granite claddings owing to the similarity between their coefficients of thermal expansion, being $10.0 \times 10^{-6}$ per °C and $11.0 \times 10^{-6}$ per °C respectively.

Where limestone, marble and granite are used in very thin sections they can exhibit an irreversible thermal expansion which may reduce their compressive strength by up to 50%. It is considered unlikely that the strength of the granites would reduce by this amount.

There are a range of defects associated with the metalwork incorporated into stonework and stone claddings. In general these are concerned with the results of corrosion of ferrous metalwork or an inability of the fixings to accommodate thermal, moisture and load-related movement. Although many stones, e.g. granite and marble, have virtually no moisture movement, this is not the case for limestone and sandstone. Unfortunately the degree of movement varies in relation to the degree of moisture absorption, material composition and direction of measurement relative to the structure of the stone.

Hard, dense and therefore relatively impervious mortars tend to concentrate water evaporation from the surface of the stone that they bind. This may increase the rate of deterioration of the stone due to chemical and mechanical factors.

### Revision notes

- Although igneous stone may be hard and durable, some types have significant moisture movement characteristics, causing a moisture movement coefficient of 0.1% to 0.3% in concrete.
- The porosity of limestone can vary between 1% and 40%. Durability can be more accurately assessed by quantifying the percentage of pores below 5 microns where $< 30\%$ = durable, $> 90\%$ = not durable.
- BRE has devised permitted exposure zones for limestones.
- Limestone can deteriorate when exposed to acidic rainwater or any other sulphurous source. The sulphurous acids can combine with the calcium carbonate to produce soluble calcium sulphate and hygroscopic calcium nitrate in the surface region of the stone.

- Water run-off from limestone can cause deterioration of sandstone by carrying calcium sulphate and calcium nitrate onto the sandstone surface. This can cause contour scaling.
- Porosity of sandstones can vary between 1% and 25% and saturation coefficients between 0.5 and 0.7. Therefore the weathering performance and frost resistance of the stone may vary even on the same building.
- Sandstones may also be damaged because of crystallization of calcium sulphate or calcium nitrate within the stone and below the surface.
- Metamorphic stone of low compaction, e.g. shales, may exhibit irreversible moisture movement when cleavage planes are disrupted.
- Where stone is used in thin sections for external cladding they must be tolerant of a wide temperature range and occasionally rapid temperature variation.
- Metal fixings associated with stonework can fail because of corrosion, or an inability of the fixings to accommodate thermal, moisture and load-related movement.

## ■ Discussion topics

- Compare the influence of differential moisture and thermal movement on the durability of two types of stone used in construction.
- Explain the chemical deterioration of limestone, making reference to the influence of the surrounding structure.
- Discuss the limitations of using porosity as a measure of the degree of exposure permitted for limestones.
- Explain why certain types of shale may undergo irreversible expansion whereas slates are considered to have no moisture movement.

## Further reading

BRE (1964) *Design and Appearance Part 1*, Digest 46, Building Research Establishment, HMSO.

BRE (1964) *Design and Appearance, Part 2*, Digest 46, Building Research Establishment, HMSO.

BRE (1982) *The Selection of Natural Building Stone*, Digest 269, Building Research Establishment, HMSO.

BRE (1991) *The Weathering of Natural Building Stone*, Building Report 62, Building Research Establisment, HMSO.

Cook, G.K. and Hinks, A.J. (1992) *Appraising Building Defects: Perspectives on Stability and Hygrothermal Performance*, Longman Scientific & Technical, London.

Everett, A. (1975) *Materials – Mitchells Building Construction*, Longman Scientific & Technical, London.

PSA (1989) *Defects in Buildings*, HMSO.

Ransom, W.H. (1981) *Building Failures: Diagnosis and Avoidance*, 2nd edn, E. & F.N. Spon, London.

Richardson, B.A. (1991) *Defects and Deterioration in Buildings*, E. & F.N. Spon, London.

Taylor, G.D. (1991) *Construction Materials*, Longman Scientific & Technical, London.

# 1.3 Aggregate problems generally

## Learning objectives

You should be aware of:

- the potential effect of aggregate problems on the surface quality, integrity and strength of concrete;
- the range of symptoms of such problems.

Aggregate form, grading and deposition have a central impact on the quality, durability and strength of concrete. Movement in coarse aggregate is usually negligible, since the problem aggregates have largely been eliminated. Obviously, any aggregate shrinkage will have profound consequences for the concrete, and for shrinkage failure.

Problem aggregates such as basaltic and doleritic stone may produce a moisture movement coefficient in the order of 0.1% in the concrete. Cracking will occur and, in the case of fine aggregate, shrinkage will take the form of map cracking in the cement paste. Differential shrinking stresses may induce bending in asymmetrical components, and in heavily reinforced sections the cracking will appear at the relatively weak changes in section. In extreme cases the concrete may disintegrate completely or be rendered more vulnerable to frost attack or reinforcement corrosion.

Problems may also arise with aggregates containing excessive amounts of chalk, clay, coal, sulphates or organic material. Aside from staining or pop-out, which affect the surface quality, there can be serious structural problems for concrete with some aggregates.

Chalk content can lead to patchy weakness and absorbency, and to localized pop-outs in the concrete in frosty conditions. Clays in aggregates can create a contaminant film on the aggregate which leads to weak planes in the concrete and weak internal bond generally. The result can be a relatively weak concrete with reduced abrasion resistance and impermeability. This latter fault is the consequence of the clay in the mix significantly increasing the water needed for a given workability. The hardened concrete is left highly permeable.

Chlorides have been used in concrete as accelerators. In plain concrete this usually causes minimal problems. However, chloride attack on steel reinforcing can be problematic and is covered in section 2.3.

*The Technology of Building Defects.* Dr John Hinks and Dr Geoff Cook.
Published in 1997 by E & FN Spon, 2–6 Boundary Row, London SE1 6HN, UK. ISBN 0 419 19770 2

The presence of coal produces a range of problems depending on the type of coal. Black staining of anthracite can occur, whereas the bituminous and lignite coals can exude tarry matter and stain. Frost sensitivity may also be a problem. Coals may also contain sulphates which attack the gel.

There is also a specific problem with alkali-reactive minerals – alkali–silicate reaction. This is discussed in section 1.4.

## Revision notes

- Aggregate form, grading and deposition have an impact on the quality, durability, and strength of concrete.
- Movement in non-problematic aggregates is usually minimal.
- Problematic aggregates include basaltic and doleritic stone.
- Symptoms of aggregate instability include cracking, which may appear as map cracking.
- Concrete may disintegrate in severe cases (rare).

---

## ■ Discussion topics

- Discuss the influence of chemical composition on the durability of natural aggregates.
- Compare the dimensional stability of dense natural aggregates with lightweight aggregates and identify the factors which influence the differences.

---

## Further reading

BRE (1987) *Concrete*, Part 1: *Materials*; Digest 325, Building Research Establishment, HMSO.

BRS (1965) *Protection Against Corrosion of Reinforcing Steel in Concrete*, Digest 59, Building Research Station, HMSO.

Honeybourne, D.B. (1971) *Changes in the Appearance of Concrete on Exposure*, Digest 126, Building Research Station, HMSO.

Shirley, D.E. (1981) *Impurities in Concreting Aggregates*, Construction Guide, Cement and Concrete Association.

# 1.4 Alkali–silicate reaction

**Learning objectives**

- A relatively rare defect which you should be aware of
- Distinctions between this and other causes of map cracking are important.

Reactive silicates which are present in some aggregates can react with the alkaline sodium or potassium oxides produced by the hydration of ordinary Portland cement. This is usually termed alkali–silicate reaction (ASR), but may be known as the alkali–aggregate reaction. It creates a gel around the aggregate, producing an increase in its volume. The effect is uncommon in the UK. The susceptible UK aggregates are of the siliceous type, for example opal.

The mechanism requires a high moisture level in the concrete, and wetting/drying cycles will accelerate the process. It is usually the fine aggregate that reacts (with the alkaline pore fluid in the concrete), producing a calcium alkali silicate gel at its surface. As this gel absorbs moisture it swells, possibly causing the concrete to crack. Affected concrete suffers a loss of integrity and strength.

In constrained concrete the expansion cracking will create a relatively ordered pattern. In unrestrained concrete a network of fine, random cracking appears, bounded by fewer, larger cracks. This is termed map cracking, and can take on an appearance similar to that of shrinkage or frost-related cracking. Alkali–silicate cracking develops slowly, sometimes taking as long as five years to appear. In contrast, drying shrinkage occurs within the first year or so of casting and frost attack may be distinguishable where it produces surface spalling of the concrete. Note, however, that frost attack may be a secondary defect following spalling produced by alkali–silicate reaction-induced expansion.

Alkali–silicate reaction can be confirmed by microscopic analysis. In severe cases, the colourless jelly produced by the reactions may be visible. Research findings suggest that the damage to concrete affected by ASR is generally confined to 25–150 mm thickness of concrete. For many building applications this is still a significant depth.

*The Technology of Building Defects.* Dr John Hinks and Dr Geoff Cook.
Published in 1997 by E & FN Spon, 2–6 Boundary Row, London SE1 6HN, UK. ISBN 0 419 19770 2

## Revision notes

- A reaction between alkaline sodium or potassium oxides produced by the hydration of ordinary Portland cement and reactive silicates present in aggregate.
- Reaction produces clear gel which absorbs water and cracks concrete as it swells.
- Rare in UK aggregates.

---

## ■ Discussion topics

Care must be exercised when identifying an alkali–silicate reaction, since the symptoms of alkali–silicate reaction may be similar to drying shrinkage and frost attack. Explain this in detail, and how you would distinguish between ASR and other causes of cracking.

---

## Further reading

Bonshor, R.B. and Bonshor, L.L. (1996) *Cracking in Buildings*. Construction Research Communications (BRE), pp. 12, 67.

BRE (1991) *Alkali Aggregate Reactions in Concrete*, Digest 330, Building Research Establishment, HMSO (revised 1991)

Chana, P.S. (1989) Bond Strength of Reinforcement in Concrete Affected by Alkali Silicate Reaction. *Transport and Road Research Laboratory Contractor Report No 141.*

Richardson, B.A. (1991) *Defects and Deterioration in Buildings*, E. & F.N. Spon, p. 123.

Somerville, G. (1985) Engineering Aspects of Alkali–Silica Reaction. *Cement and Concrete Association Interim Technical Note 8.*

Walton, P. (1993) *Effects of Alkali–Silicate Reaction on Concrete Foundations*, BRE Information Paper IP 16/93.

Williams, G.T. (1982) Basic facts about concrete II – faults. *Structural Survey* **1**(2), 170–175 (Summer).

# 1.5 Adhesives

## Learning objectives

The student should be aware of the different types of adhesive used in the construction industry. You should be able to:

- describe the characteristics of different types of adhesive used in the construction industry;
- describe the mechanisms associated with defects in the application and use of adhesives in the construction process;
- discuss the issues associated with the use and potential failure of adhesives in the construction industry.

## Types and classification

There are, and have been, many different types of adhesive in general use in the construction industry. The adhesive will bond itself together when correctly mixed, applied and cured. It will only bond to the background if it is compatible with this background and is in close proximity to the materials of the background. The traditional adhesives get within the surface of the background and then set through an 'interlocking' of the adhesive within the material rather than bonding to it. This action can be described as cohesive and not adhesive. The relatively new group of adhesives, including the epoxy resins, are capable of bonding to impervious surfaces.

Adhesives may be classified in accordance with their durability as:

- WPB = weather- and boilproof types
- BR = boil resistant
- MR = moisture resistant
- Int = resistant to cold water.

Types of adhesive include the following.

## Organic adhesives

- Animal glues (Int). These are specified in BS 745. They are not resistant to water.

*The Technology of Building Defects.* Dr John Hinks and Dr Geoff Cook.
Published in 1997 by E & FN Spon, 2–6 Boundary Row, London SE1 6HN, UK. ISBN 0 419 19770 2

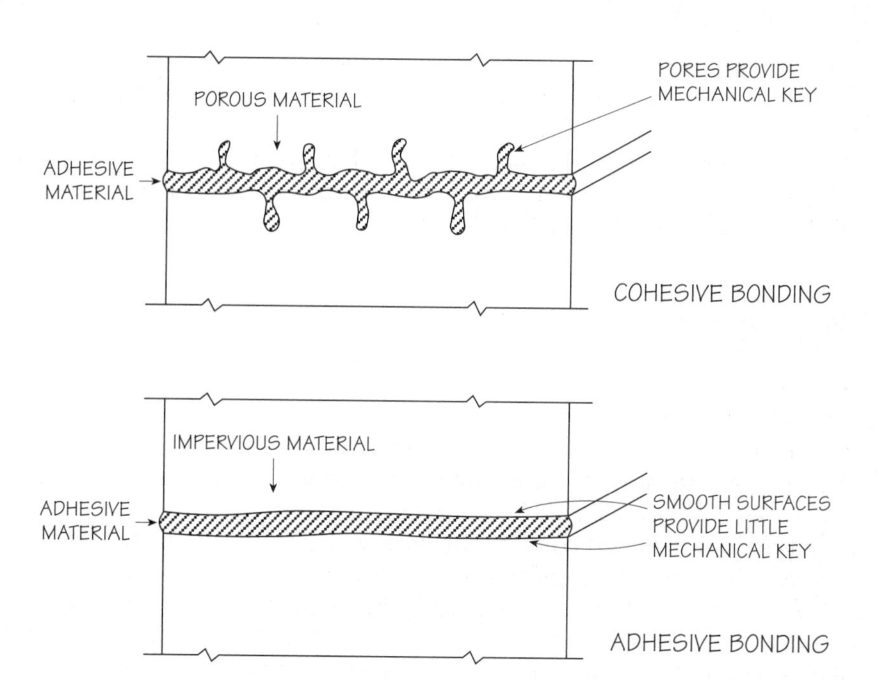


**Fig. 1.2.** Adhesive and cohesive bonding.

- Casein glues (Int). These are derived from skimmed milk and since they can be mixed cold, have a 6 hour pot life. Generally suitable for interior applications.

Both of these adhesives are mixed in water and require either evaporation or absorption of this water to occur in the curing process.

## Thermosetting adhesives

- Urea formaldehyde (MR). This is available in a powder or two-pack form. Can be used on porous surfaces and there is a need for the application of pressure during curing. Crazing can occur with large volumes of adhesive. Not hot-water resistant. Urea formaldehyde adhesive used in external doors can fail when wet. Melamine–formaldehyde (BR) adhesives cure at 100°C and can be used where urea formaldehyde types are not suitable.
- Phenolic resins (WPB) have excellent moisture resistance and strength. They may be brittle and generally require factory conditions for application.
- Resorcinol formaldehyde (WPB) adhesives cold-cure with formaldehyde. These can be used for gluing timber which will be used in external applications. Timber will still require treatment. They can be used where backgrounds are porous, e.g. brickwork.
- Epoxy resins (WPB) are claimed to be able to bond anything. The two-part adhesives are common. Considerable weather resistance and excellent mechanical strength. Is seen as a key adhesive in the development of glued joints in structural timber.

- Alkyl cyanoacrylate types are the so-called 'super-glues'. These have almost instantaneous bond strength, and can bond together fingers. Surgery may be required for separation.
- Certain types of polyurethane (better than MR) adhesives may emit formaldehyde vapour which may be a source of discomfort. There is some controversy concerning the health risk of formaldehyde.

## Thermoplastic adhesives

These are generally more prone to creep than the thermosetting types. This can result in joint movement.

Polyvinyl acetate (Int) is water miscible. Mainly used on interior timber since the adhesive softens on wetting. One surface must be absorbent. It is also used as a bonding agent for concrete.

Polystyrene adhesives contain solvents which are flammable. They use the solution principle to join PVC and polystyrene.

The water-based bituminous adhesives have low tensile stress resistance. Commonly used as a PVC floor tile adhesive. This bond can fail owing to moisture migration through the screed, since the moisture breaks down the bond between screed and adhesive. It is possible for the adhesive to be firmly attached to the tile. Sodium carbonate may be left as a white mark between and around the tiles. Where there is a long time between laying the adhesive and laying the tiles, the bond may be permanently weakened.

Many of the rubber-based types are contact adhesives. The two surfaces are coated with adhesive and left for the solvent to evaporate. On contact the surfaces adhere. Positioning is critical and can be difficult to modify.

With the wide range of adhesives, it is not surprising that some have failed or have been used incorrectly. This may be due to a range of factors, including:

- incorrect storage, where temperature is a key factor, or they may have been used after their shelf life;
- application in conditions outside the recommended temperature and humidity range, which may also affect the long-term performance of the adhesive.

Two-part adhesives may be mixed incorrectly. Too much hardener will reduce curing time and may result in a brittle joint. Too little hardener may mean extended curing times and reduced long-term strength.

Insufficient preparation of surfaces may fail to provide an acceptable degree of mechanical key. Where surfaces are insufficiently clean, this may act as a barrier between the adhesive and the surface to be bonded.

The adhesive may be incompatible with the surfaces. This can be a particular problem with the solvent welding compounds used with uPVC pipework since each material can have a different chemical composition.

Organic-type adhesives, including epoxy or polyester resins or mortars containing styrene butadiene, are strong and durable. These can be used to fix brick slips to concrete, although they can fail because of differential moisture and/or thermal movement.

## Revision notes

- There is a need for adequate surface preparation, adhesive mixing (where required), coverage and curing conditions for an effective adhesive joint to be formed.
- There are many different types of adhesive, each having different characteristics. Water-based types may require absorbent backgrounds for effective curing.
- Animal-based adhesives are generally only suitable for interior applications.
- The thermosetting types offer greater joint rigidity than the thermoplastic types.
- The bond between background and water-based adhesives can be broken down by moisture migration through the background.
- Epoxy resins are claimed to be able to bond anything, even non-absorbent materials.

---

## Discussion topics

- Discuss the characteristics of organic and inorganic adhesives in relation to their reliability and durability.
- Taking examples from the construction industry, compare the application and performance of adhesives in a factory environment with those used on site.
- Explain the failure mechanisms in thermosetting and thermoplastic adhesives when joining timber.
- Compare the construction and performance of a mechanical joint with a joint formed with an adhesive.

---

## Further reading

BRE (1964) *Sheet and Tile Flooring made from Thermoplastic Binders*, Digest 33, Building Research Establishment, HMSO.

BRE (1986) *Gluing Wood Successfully*, Digest 314, Building Research Establishment, HMSO.

BRE (1989) *Choosing Wood Adhesives*, Digest 340, Building Research Establishment, HMSO.

Everett, A. (1975) *Materials – Mitchells Building Construction*, Longman Scientific & Technical, London.

Taylor, G.D. (1991) *Construction Materials*, Longman Scientific & Technical, London.

# 1.6 Glass

### Learning objectives

You should be able to:

- describe the general composition and properties of glass used in construction;
- describe the environmental deficiency factors which affect glass in construction;
- identify the thermally induced deficiency factors associated with glass;
- describe the mechanical deficiency factors which affect glass in construction;
- be able to discuss the deficiency issues associated with the use of glass in construction.

## General

There are many different types of glass used in construction. Laminated types can be shatterproof, impact resistant and thermally insulating. Most glasses are of the soda-lime type made from silicon oxide ($SiO_2$), sodium carbonate ($Na_2CO_3$) and calcium carbonate ($CaCO_3$). Glass has an amorphous structure and a tendency to form surface cracks. These cracks have an effect on the tensile strength of the glass. For example, thin fibres of flawless glass can have a tensile stress at failure of 300 /mm$^2$, whereas bulk glass containing flaws may fail at stresses $< 100$ N/mm$^2$.

Glass has significant chemical resistance, can be considered as impermeable and therefore is considered to be highly durable. Unfortunately there are a range of environmentally and mechanically induced factors which can cause the deterioration of glass.

## Environmental factors

The importance of energy conservation in construction has focused attention on the type of glazing in windows and rooflights. Double glazing is now

*The Technology of Building Defects.* Dr John Hinks and Dr Geoff Cook.
Published in 1997 by E & FN Spon, 2–6 Boundary Row, London SE1 6HN, UK. ISBN 0 419 19770 2

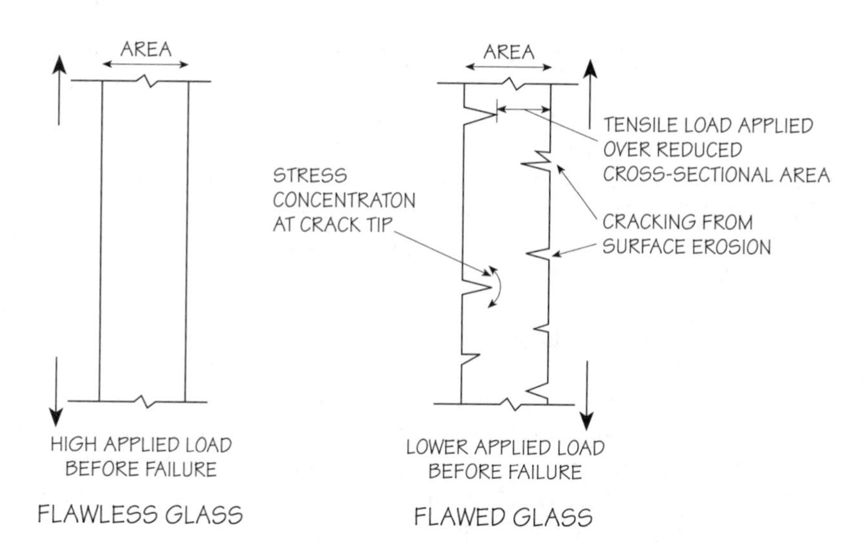


**Fig. 1.3.** Effects of flaws on the strength of glass.

commonplace. The thermal conductivity of glass is around 1.1 W/m K, and this compares with a typical thermal conductivity of timber of 0.14 W/m K. The advantage of double glazing for thermal insulation is to utilize the low thermal conductivity of air, typically 0.025 W/m K, and the surface resistances of the four glass surfaces. It is common for the air cavity and the inner surface resistances of the glass to be treated as an overall thermal resistance when calculating heat flow. It is suggested that single glazing will provide a '$U$' value of 5.6 W/m$^2$K and double glazing 2.8 W/m$^2$K. Where metal window frames without thermal breaks are used they have a greater thermal conductivity than timber, typically 164 W/m$^2$K for aluminium and 43 W/m$^2$K for steel. This will affect the overall heat loss through the window and since the inner surface temperatures of the metal frames are likely to be lower than timber under steady-state heat loss conditions, there is an increased risk of condensation.

Whilst there is a need to reduce heat flow from buildings, there is also a need to reduce solar gains into some buildings. This can be achieved by changing the emissivity of the glass. Whilst a typical emissivity for glass and many construction materials is about 0.9, that of a low-emissivity glass may be less than 0.5. Since by reducing the emissivity of the glass surface the surface resistance is increased, this will cause the heat flow through the glass to be reduced. This generally operates to reduce heat flow from the building. Thin metallic films are commonly used and, although effective thermally, they can reduce the light transmission of the glass. Single glazing has a light transmission coefficient of around 0.8; this compares to 0.6 for some low-emissivity glass. This light loss can increase where poor cleaning and maintenance regimes apply. Since the metallic films are usually tinted they provide the occupants with a permanently coloured outside view. The thin metallic films can also be used on the cavity side of the inner glass to reduce heat loss from the building, although they have the ability to transmit the shorter-wavelength solar gain.

An alternative approach is to use heat-absorbing glass. These are commonly coloured and although they reradiate some of the absorbed heat the light transmission coefficients can be as low as 0.5. With approximately 50% of the solar heat absorbed by the glass, then failure can occur where there is inadequate expansion and contraction provision. This can be difficult where differential temperatures exist across the glass sheet as when part of a window or glazed cladding is shaded. Indeed, thermal movement of glass within frames is a general problem. Typically metals have a coefficient of thermal expansion about twice that of glass. For example that of mild steel is $11 \times 10^{-6}$ per °C and glass $7 \times 10^{-6}$. Therefore insufficient clearance between glass and frame or an inability of the glass to move can cause cracking. Movement provision for heat-absorbing glass is particularly important.

Many of the cavities in double-glazed units are sealed. Failure of the cavity seal may allow moisture and dirt to enter and may cause condensation and/or mould growth to occur. This can increase heat loss, reduce light transmission and look unsightly. A ventilated cavity may reduce the risk of condensation but increase heat loss.

The typical 20 mm maximum double-glazed cavity width required for thermal insulation makes only a small contribution to sound reduction. This is more likely to be achieved by cavities of around 200 mm width, which are less effective as thermal insulators. Air gaps around the joints between glass and frame, frame and structure and opening sections of windows can substantially increase the sound transmission coefficient.

## Mechanical factors

Glass can lose a significant amount of strength with age because of surface erosion. This may be due to the effects of weathering or wear and tear. Rain washing from corroded metallic surfaces can cause surface discoloration of the glass, accelerating the weathering process. When the surface becomes roughened there is an increased likelihood of surface cracks being exposed. These can cause stress concentrations around the root of the crack and cause the glass to fracture. In this way glass can become brittle with age, making it more difficult to cut and handle.

Fire can cause glass to melt or shatter. Where fire-resistant construction is required then heat resisting, intumescent coated or Georgian wired glass can be used. Since the integrity of the glazing is influenced by the frame, the beads and fixings must retain the glass in position under fire conditions. Traditionally this was achieved by the adequate size of timber beads.

The edge effects of thermal expansion and cracking are a particular problem with Georgian wired glass. Shelled or poorly cut edges are a starting point for stress cracking. The wires can corrode and cause expansive cracking of the glass, which may occupy the edge gap provided for thermal expansion.

Surface discoloration and etching of glass can occur when the glass is closely stacked under damp or wet conditions. The water enters the space between the sheets by capillary action and can also cause mould growth. Alkali paint strippers can cause etching of the glass surface as can the alkalis in cementitious mortar or concrete. This may be washed from the surrounding structure. Although resistant to many acids, glass is affected by hydrofluoric acid. Surface discoloration and etching can reduce appearance and light transmission properties. A dirty glass surface exposed to solar radiation can cause the glass surface to heat up locally and cause cracking induced by differential thermal expansion.

Glass can shatter into dangerous fragments, particularly under impact load. Since this can occur in buildings, particularly where occupants can walk into glass screens and doors, there are a range of locations where the use of single glazing is inappropriate. A variety of suitable glass offering the required level of safety is available. This safety level is based on an average-sized child running into the glass. Either the glass should resist the impact or the resulting damaged glass should present little health risk. Since this glass should be marked to indicate compliance with an acceptable standard of performance, this should form part of any building survey of appraisal.

## Revision notes

- Soda-lime glass has an amorphous structure and a tendency to form surface cracks. These cracks reduce the tensile strength from that of flawless glass.
- Double glazing provides a 'U' value of 2.8 $W/m^2 K$, compared to 5.6 $W/m^2 K$ for single glazing.
- Low-emissivity glass in a double-glazed unit can reduce solar gain and heat loss from within buildings although thermal movement is likely to be greater than with normal glass.
- Tinted or coloured glass can reduce light transmission and may increase heat absorption.
- Cracking can occur where there is inadequate thermal movement provision around the glass.
- Failure of a double-glazed cavity seal may allow moisture and dirt to enter, causing condensation, mould growth and increased heat loss to occur.
- Small double-glazed air cavities are unsuitable for reducing sound transmission and large cavities are unsuitable for reducing heat loss.
- Surface erosion reduces tensile strength and may affect light transmission and thermal movement.
- Fire can cause glass to melt or shatter.
- The metal wires in Georgian wired glass can corrode causing cracking and expansion.
- Impact damage to glass can produce dangerous fragments. Safety glass can be identified by markings.

## ■ Discussion topics

- Identify the features which influence deficiency in the sound-reducing effects of double glazing.
- Identify the features which influence deficiency in the thermally insulating performance of double glazing.
- Compare the influence of differential movement and long-term weathering on the failure mechanisms associated with single glazing.
- Explain how glass may fail to provide a safe environment during normal use of a building and during a fire.
- Explain what factors can influence the mechanical properties of glass.

## Further reading

BRE (1993) *Double Glazing for Heat and Sound Insulation*, Digest 379, Building Research Establishment, HMSO.

BRE (1992) *Selecting Windows by Performance*, Digest 377, Building Research Establishment, HMSO.

Cook, G.K. and Hinks, A.J. (1992) *Appraising Building Defects: Perspectives on Stability and Hygrothermal Performance*, Longman Scientific & Technical, London.

Everett, A. (1975) *Materials – Mitchells Building Construction*, Longman Scientific & Technical, London.

Garvin, S.L. and Blois-Brooke, R. (1995) Double Glazing Units: A BRE Guide to Improved Durability. *Building Research Establishment Report 280.*

PSA (1989) *Defects in Buildings*, HMSO.

Richardson, B.A. (1991) *Defects and Deterioration in Buildings*, E. & F.N. Spon, London.

Taylor, G.D. (1991) *Construction Materials*, Longman Scientific & Technical, London.

# 1.7 Plastics

### Learning objectives

You should be able to:

- compare the deficiency characteristics of several different types of plastic used in the construction industry;
- describe thermally induced effects which can cause deterioration of plastics as used in the construction industry;
- explain the influence of glass transition temperature on the mechanical properties of plastics;
- describe the effects of UV, ozone and strong alkalis on plastics;
- compare the long-term mechanical performance of plastics;
- explain the importance of identification when assessing the deterioration of plastics.

## Generic types

The differences between plastics can be generalized by the following broad, simplistic definitions.

- Thermoplastic (TP) plastics. These are generally softened when heated, and are flexible, tough and mouldable.
- Thermosetting (TS) plastics. These are not generally softened when heated, and are strong, stiff and relatively heat resistant.

Most plastics appear to be reasonably durable, with some having Agrement certificates which estimate lives of 30 years or more. The uPVC window could be considered to be working reasonably well at the present time, although like its timber alternative it is combustible.

## Thermally induced effects

The glass transition temperature ($T_g$) identifies a point at which the physical properties of plastics change. Below $T_g$ the plastic can be considered brittle and solid, whilst above $T_g$ it can be considered a viscous liquid. There is, unsurprisingly, an intermediate state somewhere between the two. This is

*The Technology of Building Defects.* Dr John Hinks and Dr Geoff Cook.
Published in 1997 by E & FN Spon, 2–6 Boundary Row, London SE1 6HN, UK. ISBN 0 419 19770 2

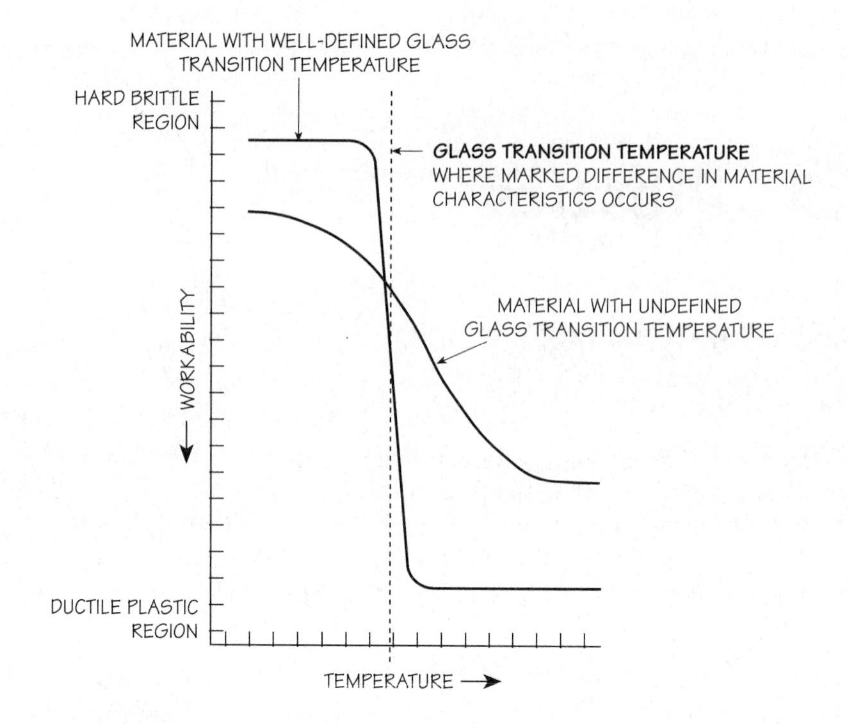


**Fig. 1.4.** The general physical property changes associated with the glass transition temperature zone.

the region where elastomeric properties exist. These materials are used for movement joints, isolation and bridge bearings. Since only crystalline materials exhibit well-defined melting points it follows that the degree of crystallinity of plastics varies considerably. Where plastics are used outside their recommended temperature range, then their physical properties will no longer match the performance requirements. A brittle or plastic failure may occur.

Plastics generally have high coefficients of expansion. This can cause pipes and gutters to leak and other components to distort. In the case of roof sheeting there can be a stress build-up around the fixings causing sheets to tear. Pipework requires adequate support or joints, since many types are a push fit and will leak. Plastics tend to exhibit time-dependent strain, or creep. Under a high stress this is more pronounced and can be reduced by additional support or restraint.

## General deficiency factors

Ultraviolet degradation can occur with certain types of plastics. This is a particular problem near coasts, although a range of additives to the basic plastic compound claim to reduce the long-term effect. Unfortunately some may also reduce the fire resistance of the plastic. Ozone is also presumed to degrade plastics. This can also be more prevalent near coasts although it is considered to be a minor problem.

There are a considerable number of plastics used for above-ground drainage fittings and pipework. Some of these can be affected by chemical waste, which would include domestic paint strippers.

Since the nature of defects in plastics is specific to the particular type of plastic, correct identification is particularly important. The BRE have produced a kit to help identification. This includes samples and guidance concerning simple tests. The Rubber and Plastics Advisory Service can also recommend specialists and testing laboratories.

A selection of plastics, and their potential for deficiency, includes the following.

## Unplasticized polyvinyl chloride (uPVC) (TP)

Accounts for around 50% of total plastics in the construction industry. Used for above-ground drainage, wall cladding, translucent or opaque corrugated roof sheeting, ducting, skirting and architraves. Also for window frames when reinforced with metal, although even then distortion may occur. High coefficient of thermal expansion. Can be solvent welded. Painting may adversely affect impact resistance. Translucent uPVC rooflights may be more brittle than GRP types and can be damaged by hail.

## Plasticized polyvinyl chloride (PVC) (TP)

Used for floor covering, sarking felt, as a membrane covering for flat roofs and for water bars. Expanded PVC is used as an insulation material. PVC has a high coefficient of thermal expansion. Some adhesives used for PVC floor tiles can cause the plasticizer in the floor tile to migrate into the adhesive. This softens the adhesive and allows the tiles to move under traffic, causing the joints to open.

## Polyvinyl fluoride (PVF) (TP)

This has a tendency to crystallize with resulting increases in stiffness. Used as a coating material for metal and timber.

## Polyethylene (TP)

Used for cold-water cisterns, although oil-based jointing compounds may cause cracking. Used also for ball valve floats, cold-water pipes, above-ground drainage pipes and fittings, damp-proof membranes and electrical conductor insulation. Some types have substantial coefficients of thermal expansion. Although polyethylene is relatively unaffected by acids and alkalis it has a slightly greasy surface which makes painting difficult. Adhesives are equally difficult to apply and can fail where there is inadequate surface roughening to provide mechanical key.

## Acrylic resins (TP)

Used for sinks, draining boards, baths, corrugated sheeting and rooflights. Polymethyl methacrylate, commonly called Perspex, has largely replaced cast iron for baths. It can be marked by burns and scratches. Abrasive cleaning agents can remove the shiny surface.

## Acrylonitrile butadiene styrene (ABS) copolymers (TP)

Used for large drainage inspection chambers. ABS softens at a higher temperature than uPVC and therefore is more suitable for above-ground drainage systems. Solvent welding for ABS must be carried out with the correct solvent; PVC solvents may give an initial appearance of being effective but they may not have any solvent action on ABS.

## Polypropylene (TP)

Used for plumbing and drainage fittings, wall ties and fibre reinforcement. Whilst capable of tolerating high temperatures, $>80°C$, it exhibits considerable expansion.

## Polycarbonates (TP)

Used for vandal-resistant glazing. Has a high coefficient of expansion. The material becomes increasingly opaque with age, adversely affecting the light transmission coefficient. This is a particular problem when used in exposed positions. The material melts away from flames, although not readily supporting combustion.

## Polystyrene (TP)

Expanded for insulation, since it has a very low thermal conductivity, 0.033 W/m°C. Unmodified polystyrene can burn readily.

## Phenol formaldehyde resins (bakelite) (TS)

This was one of the first forms of commercially available plastic. Phenol formaldehyde resins are impregnated into paper and fabric to make roof and wall sheeting, and expanded for insulation material.

## Polyester resins (TS)

These can be reinforced with glass fibre to produce glass-reinforced plastic (GRP). This has been for external cladding, cold water cisterns, storage tanks and large gutters. It has a high coefficient of expansion. Moisture can reduce the bond strength between glass fibres and the polyester resin. The moisture source may be from condensation where internal surfaces are affected. General weathering can roughen the surface and allow dirt to build up.

## Epoxide resins (TS)

Used for in-situ flooring and concrete repair. Where there is a difference between the physical and chemical performance characteristics of the epoxide resin and the concrete then deterioration of the repair may occur.

## Polyurethane (TS)

This can be expanded for use as thermal insulation or used unexpanded in paints. Although the paint films have the useful properties of hardness and strength, where there is a need to accommodate thermal or moisture-induced movement the paint film may fail.

## Urea formaldehyde (TS)

Used in an expanded form for thermal insulation. Formaldehyde may cause toxicity by inhalation. Urea formaldehyde contains unreacted formaldehyde which is slowly released. Formaldehyde has been associated with allegations concerning the cause of illness and allergic reaction in building occupants. There appears to be no evidence that formaldehyde is carcinogenic to humans. The use of urea formaldehyde is the subject of several standards.

### Revision notes

- Plastics can be classified as thermoplastic or thermosetting plastics, with the former generally more likely to distort plastically with heat and the latter generally more crystalline and liable to a brittle failure.
- Plastics are affected by temperature changes. When above their glass transition temperature ($T_g$) they are generally viscous liquids. When below $T_g$ they are generally brittle.
- Where the generally high coefficients of expansion of plastics have not been accommodated, or adequate support provided, distortion and deterioration can occur.
- Plastics can exhibit creep under long-term loading, and be affected by UV radiation and to a lesser extent ozone.
- Chemical resistance varies with type of plastic. Identification is essential and complicated by the similar appearances of different types of plastic.

## ■ Discussion topics

- Discuss the effects on the nature of building deficiency caused by the growth in the use of plastics in construction.

- Compare the deterioration mechanisms that apply to the thermoplastics and thermosetting plastics which are used in the construction industry.
- Explain the effect of weathering on the mechanical properties of GRP and uPVC.

---

## Further reading

BRE (1974) *Reinforced Plastics Cladding Panels,* Digest 161, Building Research Establishment, HMSO.

BRE (1980) *Cavity Insulation,* Digest 236, Building Research Establishment, HMSO.

BRE (1985) *Fire Risks from Combustible Cavity Insulation,* Digest 294, Building Research Establishment, HMSO.

Everett, A. (1975) *Materials – Mitchells Building Construction,* Longman Scientific & Technical, London.

PSA (1989) *Defects in Buildings,* HMSO.

Richardson, B.A. (1991) *Defects and Deterioration in Buildings,* E. & F.N. Spon, London.

Taylor, G.D. (1991) *Construction Materials,* Longman Scientific & Technical, London.

CHAPTER 2

# Failure mechanisms in cementitious materials

# 2.1 Cement

## Learning objectives

You should be able to

- describe the characteristics and compare the causes of deficiency in high-alumina cement and several types of Portland cement;
- describe a range of properties which are tested to determine the quality of cements;
- describe the characteristics and compare the causes of deficiency in natural and artificial aggregates and mixing water;
- be able to discuss the issues associated with the deficiency of different types of cement.

## Cement generally

Ordinary Portland cement (OPC) is a type of hydraulic cement. The manufacture of Portland cements in the UK is closely controlled and can be assessed in relation to a range of British Standard specifications. Care is needed when using imported cements to assess their quality and their influence on the durability of concrete mixes.

Cement should be fine, having a specific surface not less than 275 $m^2/kg$ between 10 and about 100 microns, and even-textured. The quality of cement is strongly influenced by chemical composition and fineness. The chemical composition can be determined by Bogue equations. These determine the percentage of the four compounds of cement, $C_2S$, $C_3S$, $C_3A$ and $C_4AF$, from the percentage of the four base components in the raw materials used to manufacture the cement. This is an essential process for the manufacturers. The $C_3S$ content has increased since the 1950s and therefore the early strength and heat of hydration during the first 24 hours have increased. Greater care is therefore required with mass concrete pours since thermally induced cracking is more likely.

The British Standards for cements describe a range of tests, including:

- limits for combustible impurities;
- limits for acid-soluble impurities;
- limits for magnesium oxide (this may cause expansion, termed 'unsoundness', of the cement during and after hydration);

*The Technology of Building Defects.* Dr John Hinks and Dr Geoff Cook.
Published in 1997 by E & FN Spon, 2–6 Boundary Row, London SE1 6HN, UK. ISBN 0 419 19770 2

- limits for free lime (this can also cause 'unsoundness');
- limits for total sulphur, based on the $C_3A$ content. (this controls the amount of gypsum in the cement);
- initial setting to be not less than 45 minutes;
- final setting time to be not more than 10 hours.

These tests are not generally suitable for carrying out on site, since specialist facilities are needed. About 5% by weight of gypsum is added to retard the set, although the finer the cement the greater the need for retardation, and the greater the amount of added gypsum. Increasing the gypsum content, since it is water-soluble, may produce a porous concrete mix. In addition, finer cements show greater initial shrinkage since there is more water space around the smaller cement particles.

Cement should be stored in dry and ventilated conditions. Cement is hygroscopic and therefore will take moisture from the air. This causes the cement to 'go off', since hydration has commenced. Protection from frost, providing the cement is dry, is less critical. Hydrophobic cement, a type of OPC containing a water repellent, can be stored in damp atmospheres. However, the mixing time must be extended to break down the coating.

## Distinguishing problems with different types of OPC

### Rapid hardening

Specific surface not less than 350 $m^2/kg$. Considerable heat evolution during hydration and therefore should not be used in mass concrete. Ultra-rapid-hardening cement is also available, which will evolve even more heat during hydration.

### Sulphate-resisting

A maximum $C_3A$ content of 3.5% is required, since this is the compound which is the primary source of attack. Even concrete made with sulphate-resisting cement can be attacked by sulphates if porous. Good compaction and mix design are essential.

### Low-heat

The amount of $C_3S$ and $C_3A$ is reduced, and therefore the heat evolved during hydration. This also has increased resistance to sulphate attack. Rich mixes can generate significant amounts of heat during hydration.

## Problems with high-alumina cement (HAC)

Following the collapse of a roof over a school swimming pool in 1974, HAC was withdrawn from use in structural concrete.

This material is made in a different process from that used for OPC. The raw materials are limestone and bauxite, aluminium ore, and these are heated to 1600° C. The cement has an initial set in not less than 2 hours and a final set not more than 2 hours after the initial set. The consequent high early strength of HAC concrete, perhaps four to six times greater than OPC concrete for a w/c of 0.6, made the cement ideal for precast concrete sections. A considerable amount of heat is evolved during hydration, making HAC unsuitable for mass pours.

HAC has the potential for a high degree of chemical resistance and the ability to withstand high temperatures. Unfortunately, HAC concrete is subject to severe strength and durability loss under warm and damp conditions. The calcium aluminate decahydrate, the main strength-giving compound, 'converts' to tricalcium aluminate hexahydrate and aluminium hydroxide. These compounds are relatively porous, which increases the permeability of the concrete. This is particularly important since chemical resistance and strength are also reduced for converted HAC.

Even under normal temperature and environmental conditions HAC converts, although the conditions prevailing during the first 24 hours of the HAC concrete mix are critical. Overheating during this time can cause rapid conversion.

The amount of conversion can be assessed by differential thermal analysis (DTA) or testing core samples. Pull-out tests can also be used. When the characteristic 1 day cube strength is assumed to comply with CP 116, which was the recommended standard at the time that HAC was banned from structural use, then the 20 year characteristic strength is assumed to be around 20 $N/mm^2$. This allows the loadbearing capacity of HAC structural components in buildings to be assessed.

## The role of water in cement problems

The general rule that water fit for drinking is also acceptable for making concrete applies for all cementitious mixes. Sea water should not be used for concrete that is reinforced, since the normal salt concentrations can cause problems. There is also an increased risk of efflorescence when using sea water.

Water cement ratio (w/c) is a major indicator of concrete strength and therefore durability. A high w/c of 0.7 and above produces permanently porous concrete. Low w/c of 0.3 and below means that full compaction is required to achieve long-term strength potential. The variability of w/c, and therefore the variability of strength and durability, is likely to be more pronounced with site-mixed concrete.

## Aggregate-related problems

Aggregates may come from a variety of sources. This can be divided into two main groups, natural and artificial.

The natural types are formed from rocks and include the following.

### Igneous

The rock is broken into gravel and worn smooth by geological events. The material is hard, durable, impermeable and inert. Some basalts and dolerites have significant moisture movement characteristics, causing a moisture movement coefficient of 0.1% to 0.3% in the concrete.

### Sedimentary

Limestone, a very common aggregate, is a sedimentary rock. This provides the concrete mix with a low coefficient of thermal expansion and has an excellent fire resistance. Abrasion resistance is reduced. Sandstone is also a sedimentary rock. This, particularly in rocks which have not been subjected to long-term pressure during their formation, can exhibit moisture movement. This is particularly marked in the mudstones and greywackes. There can also be well-defined cleavage planes.

### Metamorphic

Rocks which have undergone modification by heat and pressure can provide hard, durable, attractive aggregates, e.g. marble.

## Artificial aggregates

### Lightweight aggregates

The absorption rates of some types may be high and variable; this in turn can lead to high moisture contents as the voids within the aggregate fill with water. Since there is an increased need to correct the w/c at the mixer there is also the risk of variation in the actual w/c used in the mix.

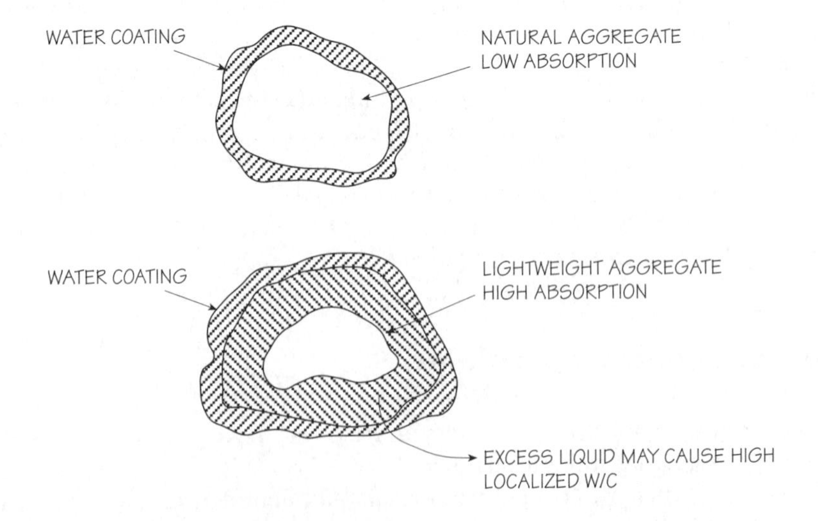


**Fig. 2.1.** Aggregate: water absorption effects.

These aggregates can have a reduced compressive stress at failure compared with natural types and this can lead to increased shrinkage. There is a greater need to allow for movement joints in structural lightweight concrete. The carbonation depth is also likely to be greater than for dense natural aggregates. Therefore reinforcement should be provided with appropriate cover depths.

### Aerated concrete

Although the reduced thermal conductivity '*k*' values are advantageous, the voids can absorb substantial quantities of water. This will increase drying times and may lead to shrinkage problems.

### No-fines concrete

This has been used for external walls. The open texture required external protection by rendering or vertical tiling. New thermal transmittance '*U*' values mean that this is unlikely to meet modern thermal insulation standards for external walls.

## Problems as a result of aggregate storage

Aggregates require storage conditions which will avoid contamination. Storage on topsoil or other materials is to be avoided and may affect subsequent concrete mixes. The stock piles should be protected from frost since this can affect the hydration and w/c ratio of subsequent concrete mixes. Where pockets of frozen aggregate exist in concrete then long-term strength can be substantially reduced.

## Grading and mixing issues

Where aggregates have a non-uniform grading there is an increased risk of voids in the concrete. Although there can be the advantage of increased workability when the 5 mm to 10 mm aggregate size is omitted, there is an increased risk of segregation. Where aggregate gradings are not consistent

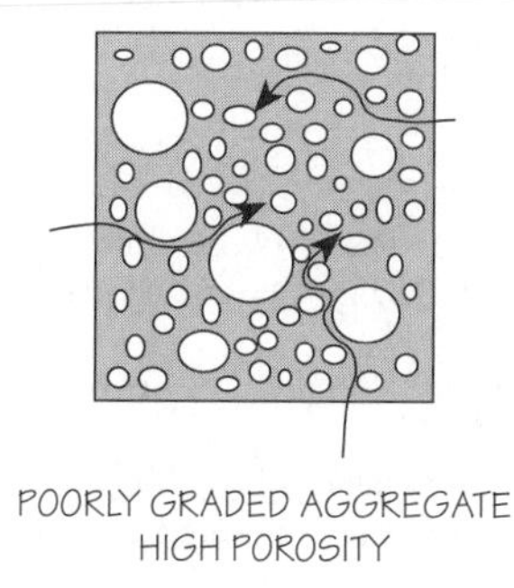


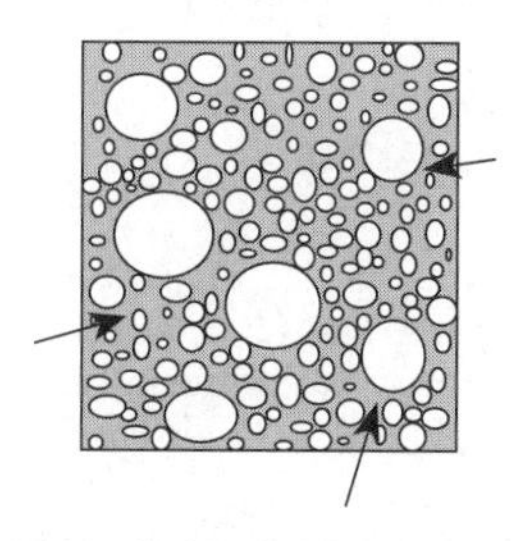


**Fig. 2.2.** Aggregate: effects of grading.

in all the concrete mixes the workability, compressive strength, quality and appearance are likely to be equally variable.

## Revision notes

- There are many types of OPCs and each has a range of potential factors which can cause deficiency. Although one type is called sulphate-resisting, high concentrations of sulphates may even attack cement made from this material.
- The fineness and chemical composition, based on the four compounds of OPC, $C_2S$, $C_3S$, $C_3A$ and $C_4AF$, are key factors when determining quality. This cannot be carried out on site.
- HAC requires carefully controlled curing conditions and can undergo conversion when used in warm, humid conditions. This produces tricalcium aluminate hexahydrate and aluminium hydroxide compounds which make the HAC more porous, increasing permeability and reducing the strength and chemical resistance of the concrete.
- Water fit to drink is likely to be acceptable for concrete mixes.
- Changes in the grading of natural aggregates can affect the consistency of concrete and may influence long-term strength. The moisture content of aggregates must be considered in relation to the w/c ratio of the mix. Certain types of artificial aggregate have the ability to absorb water and therefore possess a high moisture content. Natural aggregates can induce alkali–silica reaction.

---

## Discussion topics

- Discuss the influence of cement type on the long term durability of OPC concrete.
- Compare an inspection routine for OPC concrete structural members with that adopted for HAC concrete structural members.
- Describe the effects of changes to the aggregate type and grading on the durability of concrete.

---

## Further reading

BRE (1987) *Concrete*, Part 1: *Materials*, Digest 325, Building Research Establishment, HMSO.

BRE (1987) *Concrete* Part 2: *Specification, Design and Quality Control of Mixes*, Digest 326, Building Research Establishment, HMSO.

BRE (1988) *Alkali Aggregate Reactions in Concrete*, Digest 330, Building Research Establishment, HMSO.

BRE (1991) *Why do Buildings Crack*? Digest 361, Building Research Establishment, HMSO.

BRE (1991) *Sulphate and Acid Resistance of Concrete in the Ground*, Digest 363, Building Research Establishment, HMSO.

BRE (1994) *Assessment of Existing High Alumina Cement Concrete Construction in the UK*, Digest 392, Building Research Establishment, HMSO.

BRE (1995) *Carbonation of Concrete and its Effect on Durability*, Digest 405, Building Research Establishment, HMSO.

Cook, G.K. and Hinks, A.J. (1992) *Appraising Building Defects: Perspectives on Stability and Hygrothermal Performance*, Longman Scientific & Technical, London.

Everett, A. (1975) *Materials – Mitchells Building Construction*, Longman Scientific & Technical, London.

Neville, A.M. (1981) *Properties of Concrete*, 3rd edn. Pitman, London.

PSA (1989) *Defects in Buildings*, HMSO.

Ransom, W.H. (1981) *Building Failures: Diagnosis and Avoidance*, 2nd edn, E. & F.N. Spon, London.

Richardson, B.A. (1991) *Defects and Deterioration in Buildings*, E. & F.N. Spon.

Taylor, G.D. (1991) *Construction Materials*, Longman Scientific & Technical, London.

# 2.2 Asbestos cement

## Learning objectives

You should be able to:

- describe the different types of asbestos used in the construction industry;
- describe the characteristics of different types of asbestos used in the construction industry;
- describe some of the effects on health of inhaling asbestos fibres;
- identify construction features which may involve the use of asbestos;
- discuss the issues associated with the safe disposal of asbestos.

## Asbestos

Asbestos cement is made from the mineral chrysotile, mixed with cement and water. The amount depends on the use of the material, the greater density then the greater the cement content. Asbestos cement roofing sheets contain around 10% asbestos fibres. Roofing sheets can last up to 40 years. Asbestos fibres can be classified as follows.

### Chrysotile

This is commonly called white asbestos. This was the most commonly used mineral in asbestos cement, since it is resistant to alkalis. The fibres are relatively long and can be woven.

### Crocidolite

Commonly called blue asbestos. This type has greater strength than chrysotile and improved chemical resistance.

### Amosite

This is known as brown asbestos. The fibres are long and stiff and it was commonly used in insulating boards.

*The Technology of Building Defects.* Dr John Hinks and Dr Geoff Cook.
Published in 1997 by E & FN Spon, 2–6 Boundary Row, London SE1 6HN, UK. ISBN 0 419 19770 2

Asbestos fibres are very fine and specific surface measurements of 5000 $m^2/kg$ are possible. The specific surface for OPC is typically $< 275$ $m^2/kg$ and for rapid hardening Portland cement it is commonly $< 350$ $m^2/kg$. Asbestos has similar, small thermal expansion characteristics to those of mortars and concrete. Moisture movement is more pronounced, with up to 0.3% possible. Fixings which do not accommodate these movement cycles may be a source of failure. Algae and fungi can and do grow on asbestos cement, usually the same types which occur on concrete and cement mortar.

Carbonation will cause embrittlement of the surface, as with other cementitious materials. Carbonation also causes shrinkage and this will inter-react with the expansive effects of moisture. In this way long-term exposure results in overall shrinkage. Carbonation can therefore reduce the strength of asbestos with time, perhaps as much as a 50% reduction. Differential carbonation effects can occur on the surfaces of sheets. This may occur across the thickness of sheets or over different parts of the same side of a sheet. Either effect can cause sheet distortion.

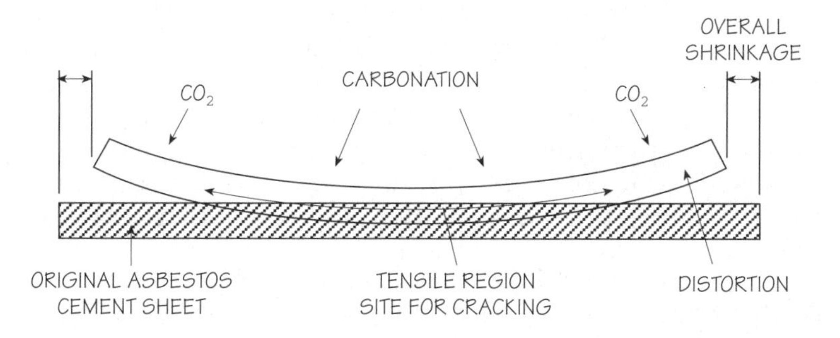


**Fig. 2.3.** Carbonation distortion of asbestos cement sheet.

Asbestosis, a disease of the lungs, is associated with exposure to asbestos. Mesothelomia and lung cancer may also be caused by the inhalation of asbestos fibres. The crocidolite, or blue, asbestos is considered particularly hazardous, and has not been imported since 1970. The loosely bound asbestos fibres can become detached owing to the effects of ageing and be carried away in the air to be subsequently inhaled. The particle size distribution in the air is an important parameter when assessing the safety of environments. A 'safe' level for chrysotile is considered to be 2 fibres, in a given size range, per millilitre of air. This reduces to 0.2 fibres, in a given size range, per millilitre of air, for crocidolite. Asbestos must be removed when fibre concentrations in the air are above prescribed limits, since this constitutes a health hazard. In other situations where asbestos is sealed, and the risk of escape into the air is virtually nil, then the decision to remove the material is less straightforward.

Asbestos may break down or shatter during a fire and this may cause the disposition of fibres over an extensive area. Where internal surfaces are lined the risk is reduced.

Great care must be exercised where it is suspected that asbestos exists. This may not be immediately apparent since it may have been used to insulate service pipework in ductwork or suspended ceilings. Fire protection of steelwork by a sprayed-on method of asbestos application was common and covered by the now withdrawn BS 3590: 1970.

The material must initially be identified and this is complicated by the similar appearance of alternative materials, particularly where they have been painted. There is also the need to determine the particular type of asbestos, e.g. blue, brown or white. An air sample and a bulk material sample must be taken. This should only be carried out by a suitably qualified person. Advice can be obtained from the environmental health department of the local authority, the Health and Safety Executive or a suitably qualified laboratory. Information can be obtained from the Asbestos Information Centre, which is funded by the Fibre Cement Manufacturers Association. The removal and disposal of asbestos from buildings is a dangerous and complex task and must only be carried out by licensed contractors.

## Revision notes

- Asbestos fibres pose a serious health hazard. This risk increases with the age of the asbestos product.
- There are generally three types of asbestos fibre: chrysotile (white asbestos), crocidolite (blue asbestos) and amosite (brown asbestos). The blue asbestos poses the greatest threat to health.
- Asbestos has been widely used in the construction industry since it possesses characteristics of durability and non-combustibility. It is therefore likely to be used in locations which utilize these properties.
- Identification is not straightforward and may require laboratory analysis. This work and any subsequent disposal must be carried out only by those who are suitably qualified to do so.

## Discussion topics

- Critically appraise the influence of asbestos-based products on the health and safety issues of building inspection.
- Produce a methodology for the inspection of buildings which recognizes the health and safety issues of hazardous materials.
- Discuss the approach to the health and safety issues of having identified asbestos in roofing felt, sprayed-on insulation to pipework and roofing sheets of an existing building.

## Further reading

Curwell, S.R. and March, C.G. (eds) (1986) *Hazardous Building Materials: A Guide to the Selection of Alternatives*, E. & F.N. Spon, London.

Department of the Environment (1983) *Asbestos Materials in Buildings*, HMSO, London.

Department of the Environment (1983) *A Guide to the Asbestos (Licensing) Regulations, HS(R) 19*, HMSO, London.

Health and Safety Executive (1980) *Asbestos*, Guidance Note MS 13, HMSO, London.

Taylor, G.D. (1991) *Construction Materials*, Longman Scientific & Technical, London.

# 2.3 Chloride attack of concrete

## Learning objectives

- You should understand the mechanism and circumstances of attack, including the sources of chlorides.
- The characteristic failure symptoms should be understood to allow you to draw the distinction between chloride attack and carbonation.

In the mid-1960s chlorides were identified as a major cause of corrosion problems in reinforced concrete. The practice of including the accelerating admixture calcium chloride in reinforced concrete was widespread in the 1970s. A frequent consequence was disruptive corrosion in the reinforcing steel, especially so where an irregular distribution of calcium chloride, as could occur with site mixing, created high concentrations of chloride ions. Where the chloride was added during manufacture, such as in the production of extra-rapid-hardening cement, the damage was somewhat minimized because of the uniform distribution and controlled concentration of the chloride. Other sources of chloride ions included the residues of hydrochloric acids used as etching media for the surface treatment of concrete cladding panels. Any unwashed residue was absorbed into the surface of the panel, producing differential concentrations across its thickness. Marine environments can produce a constant exposure to chloride attack in excess of 2 km inland, depending on the weather.

The risk of chloride attack appears to occur beyond 0.4% chloride ion by weight of cement. Below this level the normal alkalinity of concrete appears to protect against attack. Clearly any reduction in normal alkalinity, such as results from carbonation attack, will allow lower concentrations to have an effect. Above 1% by weight the risk is classified as severe, and reinforcing corrosion can occur irrespective of the normal alkaline conditions.

The attack mechanism operates as the presence of chloride ions increases the electrical conductivity of the pore water, so creating a corrosive current which accelerates the dissolution of the iron in the steel reinforcement. The result is a localized, severe pitting corrosion of the steel, leaving it notched and brittle. Failure may be sudden. The expansion of the corroding steel produces bulging or cracking of the concrete and rust staining may appear

*The Technology of Building Defects.* Dr John Hinks and Dr Geoff Cook.
Published in 1997 by E & FN Spon, 2–6 Boundary Row, London SE1 6HN, UK. ISBN 0 419 19770 2

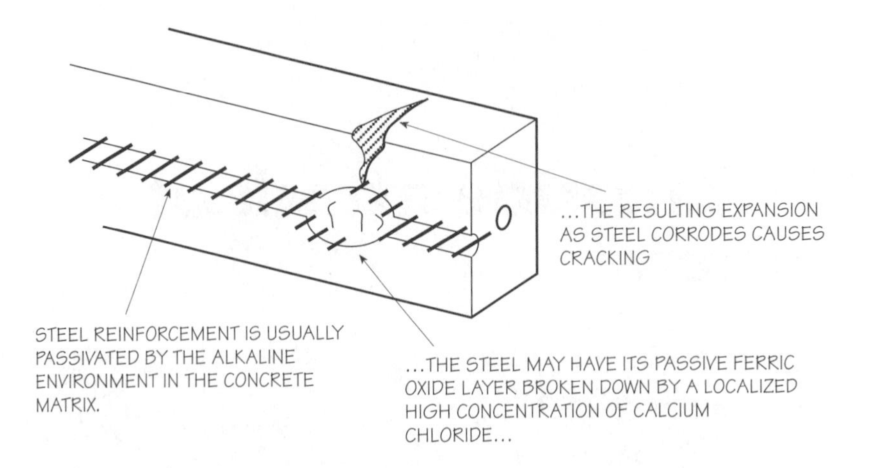


**Fig. 2.4.** Chloride attack in reinforced concrete.

at the surface of the concrete. The localized nature means that the cracking is usually isolated and wide. Chemical tests may be used to establish the presence and concentration of chloride ions in concrete.

The potential problems are not restricted to reinforced concrete. Unreinforced concrete exhibits shrinkage problems associated with the addition of significant amounts of calcium chloride, for instance a 50% increase in drying shrinkage can be produced by the addition of 0.5 to 2.0% by weight. This shrinkage increases greatly with higher concentrations of chloride ions.

## Revision notes

- A cause of major corrosion problems in reinforced concrete.
- Risk of attack occurs above 0.4% chloride ion by weight of cement. Above 1% the risk is classified as severe.
- Risk of chloride attack from EHRC less than with on-site additions of calcium chloride because of more uniform distribution of chloride ions. The addition of calcium chloride to concrete is now severely restricted.
- Presence of chloride ion increases electrical conductivity of pore water, accelerating the electrochemical dissolution of iron in reinforcement steel. Resulting corrosive current produces severe local pitting of the steel.
- Rust staining at the surface usually accompanied by isolated, wide cracks as the reinforcement is locally attacked.
- Shrinkage problems may also occur in unreinforced concrete following the addition of significant amounts of calcium chloride.
- In carbonated concrete the reduced alkalinity in the reinforcement zone leads to more severe chloride attack.

## ■ Discussion topics

- Describe the mechanism of chloride attack in reinforced concrete, and list the possible sources of chlorides. Produce a sketch diagram detailing the mechanism.
- Distinguish between the nature of attack and symptoms of chloride attack and carbonation. Explain how the tests to prove each attack would differ.

## Further reading

BRE (1982) *The Durability of Steel in Concrete*, Part 2, *Diagnosis and Assessment of Corrosion-Cracked Concrete*, Digest 264, Building Research Establishment, HMSO (August).

BRS (1965) *Protection Against Corrosion of Reinforcing Steel in Concrete*, Digest 59, Building Research Station, HMSO (June).

Richardson, B.A., (1991) *Defects and Deterioration in Buildings*, E.& F.N. Spon, London, p. 12.

# 2.4 Carbonation of concrete

## Learning objectives

- You should understand that the carbonation mechanism operates by substituting an acidic environment for the normal alkalinity of the cement matrix.
- The role of moisture and the contributory role of thin and exposed sections, shallow reinforcement cover, poor surface strength and high permeability in the progression of carbonation attacks should be clearly appreciated.
- The time feature of the appearance of cracking can be a useful distinguishing feature.
- You should also understand the principles of testing for carbonation.

Carbonation is the term used to describe the chemical reaction which occurs between alkali cement matrices, which include calcium oxide (free lime), and rainwater acidified by the absorption of carbon dioxide. The problem is usually only severe with external concrete. The resistance to carbonation is critically dependent on the permeability of the concrete, which in part will be related to its quality. Highly permeable lightweight aggregate or air-entrained concretes can be particularly vulnerable.

The reaction takes place near the surface in the pores of the concrete and produces two related problems. First, the carbonated concrete shrinks, which can further increase its porosity and hence its susceptibility to attack generally.

The second problem arises with reinforcement steel, although it may produce the first obvious symptoms. If the carbonated zone penetrates to the depth of the reinforcement steel, the acidification of the bonding concrete leads to accelerated rusting. This may be distinguishable from other causes of rusting because of its uniformity. The expansion tends to produce a hairline crack along the line of reinforcement, often extending along its whole length. Obviously, the depth of reinforcement cover is critical. Predictably, thin or highly exposed components with minimal cover suffer most.

*The Technology of Building Defects.* Dr John Hinks and Dr Geoff Cook.
Published in 1997 by E & FN Spon, 2–6 Boundary Row, London SE1 6HN, UK. ISBN 0 419 19770 2

**Fig. 2.5.** Corrosion of buried reinforcement. Carbonation, due to the poor cover of the reinforcement, is a contributing factor. (University of Reading.)

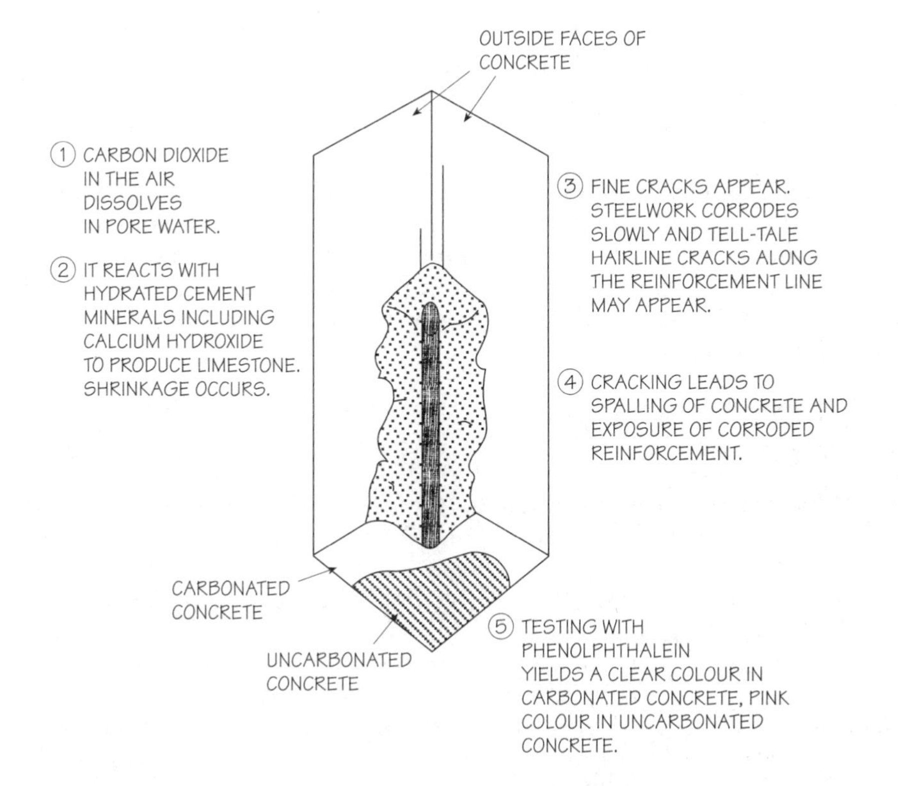


**Fig. 2.6.** The process of carbonation of concrete.

**Fig. 2.7.** The result of a phenolphthalein solution test on a fragment of concrete to establish the depth of carbonation.

It is possible for cracks caused by carbonation shrinkage to appear after only a few months of exposure. Any spalling will expose new free lime and fuel the carbonation process. In the early stages of attack the limestone by-product of the reaction tends to block the pores, so concealing and retarding the process. At this time both shrinkage and pore filling determine the rate of carbonation. Humidity is also a prime factor in the rate of carbonation. Within the intermediate ranges from 25% to 75% humidity the rate maximizes; however, if there is either too much or too little moisture carbonation is restricted. In otherwise acidic environments, such as those produced by chloride attack, the rate of carbonation penetration increases.

The depth and extent of carbonation can be confirmed with a straightforward test using manganese hydroxide or phenolphthalein solution. The reaction of phenolphthalein with uncarbonated cement leaves a pink indicator. In contrast, carbonated cement does not respond and so remains clear. A cover meter will establish the reinforcement depth across any undamaged parts of the concrete.

## Revision notes

- Process involves carbon dioxide dissolved in rainwater, so exposure is highest in built-up areas. Continues even at very low concentrations.
- The carbonation process requires moisture, making it dependent on humidity and exposure.
- Carbonation occurs mostly in the surface layer of the concrete component, so depth of attack usually progresses slowly.
- Speed of progression is slower in dense and/or low-permeability concrete.

- To some extent the process is self-limiting, and in good-quality concretes it is unlikely to extend more than 10 mm below the surface.
- The extent of reinforcement damage is increased with poor or shallow concrete cover.
- Carbonation cracks are usually less extreme than those of chloride attack, but run along the length of the component.

## Discussion topics

- Describe the process of carbonation attack on concrete, identifying the key stages in the process and the exacerbating factors. Suggest the most likely areas of a building for attack by carbonation. Produce a diagrammatic sketch of the process.
- Explain why and how thin-section reinforced-concrete components of buildings are particularly prone to carbonation attack, providing sketches to support your explanation.
- Compare and contrast the occurrence, symptoms and consequences of chloride and carbonation attack on reinforced concrete. Suggest test methods to positively distinguish between the two forms of attack. Comment on the relative scope for remedial actions.

## Further reading

Bonshor, R.B. and Bonshor, L.L. (1996) *Cracking in Buildings*, Construction Research Communications (BRE), pp. 12, 66.

Currie, R. (1986) Carbonation Depths in Structural Quality Concrete. Assessment of Evidence from Investigations of Structures and from Other Sources. *BRE Report 75.*

Honeybourne, D.B. (1971) *Changes in the Appearance of Concrete on Exposure*, Digest 126, Building Research Station, HMSO.

Neville, A.M. (1973) *Properties of Concrete*, 2nd edn, Pitman, London.

Parrot, L.J. (1987) *Review of Carbonation in Reinforced Concrete*. Cement and Concrete Association (July).

Taylor, G.D. (1983) *Materials of Construction*, Construction Press.

# 2.5 Sulphate attack of concrete and mortar

**Learning objectives**

You should be able to explain and distinguish between:

- the circumstances of sulphate attack in concrete and mortar;
- the likely symptoms of such attacks;
- the consequences for the building.

Sulphate attack affects cementitious compounds containing, or in contact with, sulphate-based materials. This includes concrete immersed in sulphated groundwater and brickwork where the mortar is affected by sulphates present in the fine aggregate or washed out of the brick clay. The basic chemical mechanisms of sulphate attack in concrete or mortar are similar, although the accompanying increases in volume produce distinctive symptoms. In general, the by-products of the chemical attack occupy a greater volume than the original cement gel and physically disrupt it, destroying its integrity and strength.

Susceptibility to sulphated groundwater attack depends on the permeability of the concrete and the porosity of the surrounding soil. The quality of waterproofing between concrete and such soils is therefore critical, and the damp-proof course (DPC) is usually a threshold of attack.

The amount and nature of sulphates present in soils are highly variable – London and Oxford Clays are problematic, as are Lower Lias and Keuper Marls. The most important naturally occurring salts are calcium sulphate, magnesium sulphate and sodium sulphate. Burnt colliery waste, once frequently used as hardcore and a light aggregate for in-situ concrete houses, residual oil shale, pulverized fuel ash and blast furnace slag can also support sulphate attack.

As the concrete absorbs the sulphated groundwater, chemical reactions occur involving calcium aluminate hydrate ($C_4A1H_3$) and tricalcium aluminate ($3CaO{\cdot}A1_2O_3$) in the hydrated cement gel. These produce a dramatic expansion (up to 227%), disrupting and spalling the concrete.

Common symptoms of sulphate attack in concrete include surface spalling and scaling. In concrete floors there is likely to be cracking and lifting of the floor slab. In addition to the disruption and possible spalling of the

*The Technology of Building Defects.* Dr John Hinks and Dr Geoff Cook.
Published in 1997 by E & FN Spon, 2–6 Boundary Row, London SE1 6HN, UK. ISBN 0 419 19770 2

concrete, there is a reduction in the alkalinity of the concrete, making any steel reinforcing more prone to corrosion. Basements and floors subject to wetting and drying cycles are generally attacked more rapidly than totally immersed substructures. Lightweight or thin sections tend to be penetrated (and hence attacked) quickly also. Sulphate attack can also arise when the free lime (calcium hydroxide) in the concrete dissolves in the water held within the pores and is redistributed throughout areas of the concrete containing sulphates.

The usual source of soluble sulphates in brickwork is the clay used for the bricks. Sodium, potassium or magnesium sulphates are common. The emergence of these salts produces relatively harmless efflorescence on exposed brickwork, which is removed by hard weather or brushing. Magnesium sulphates may crystallize just below the surface of the bricks and cause delamination.

If there is a high concentration of salts, and the presence of moisture, then sulphate attack of the mortar is likely to occur. The mortar is required to be wet for reasonably long periods if the reaction is to progress effectively, hence attack is usually limited to exposed details, such as parapets and chimneys. The sulphated solutions react with the tricalcium silicate compound in the cement mortar, producing calcium sulphoaluminate. Mortar may whiten and brickwork will swell, with individual bricks bursting.

Early symptoms include horizontal cracking in the joints between courses, as the expanding mortar creates an increase in the height of the wall. This will be reflected through any rendering, with vertical and (predominantly) horizontal cracking preceding detachment and allowing water retention, which will accelerate the sulphate attack. Total expansion may amount to 50 mm for a two-storey house, perhaps accompanied by oversailing at the corners of the building. In cavity walls the expansion usually

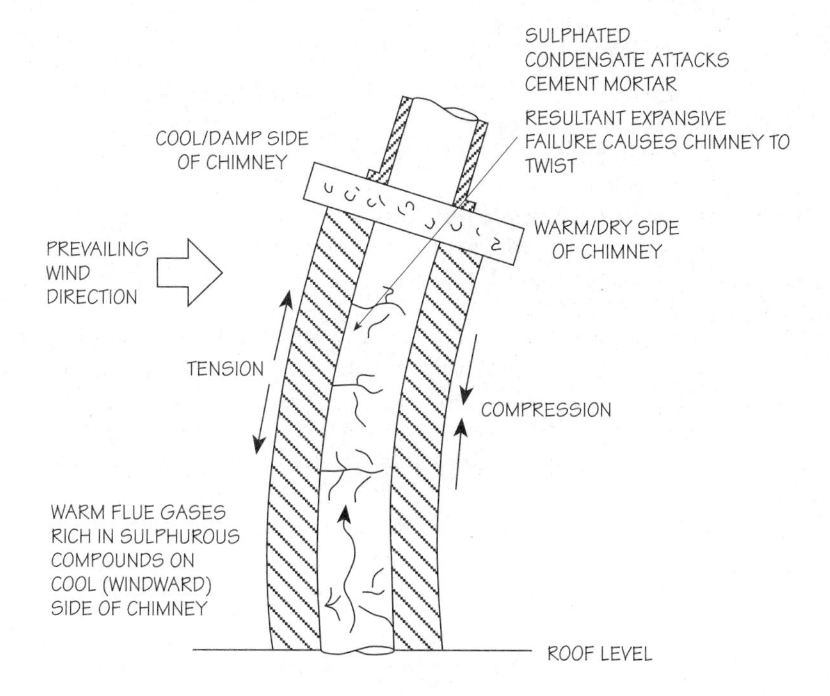


**Fig. 2.8.** Example of sulphate attack of a chimney.

**Fig. 2.9.** Cracking of cement rendering around a chimney, which may indicate sulphate attack.

only occurs in the external (damp) wall. Internal cracking at weak spots or bowing may occur as the differential expansion of the outer leaf imposes stresses on the inner leaf.

Sulphate attack usually takes at least two years to emerge, distinguishing any oversailing of brickwork from that caused by moisture expansion which occurs within the first few months following construction. The susceptibility of a material to sulphate attack may be assessed by immersion in an acidic solution and monitoring its integrity.

## Revision notes

- Process involves sulphates dissolved in water, affecting concrete below ground and mortar. Severity of attack depends on permeability of concrete and mortar, also of bricks which may be source of sulphates.
- Attack is accompanied by increases in volume causing physical disruption and loss of integrity and strength.
- Symptoms in concrete include surface spalling, scaling, cracking and distortion of lightweight components.
- Mortar may whiten and brickwork may swell, with individual bricks bursting. Early symptoms include horizontal cracking in the joints between courses, followed by vertical and predominantly horizontal cracking of external brickwork. Vertical expansion may be significant, also oversailing and possible bowing from differential expansion.

## ■ Discussion topics

- Describe the range of symptoms of sulphate attack occurring in cementitious materials, and compare the various mechanisms of such attacks. From this produce a checklist summary of the vulnerable areas of common buildings and likely consequences for the structure.
- Describe how cracking in brickwork resulting from sulphate attack may be identified and positively distinguished from other material-related causes of cracking.

## Further reading

Bonshor, R.B. and Bonshor, L.L. (1996) *Cracking in Buildings*, Construction Research Communications (BRE), pp. 10, 63–66.

BRE (1971) *Sulphate Attack on Brickwork*, Digest 89, Building Research Establishment, HMSO (January 1968 with minor revisions in 1971).

BRE (1981) *Concrete in Sulphate-Bearing Soils and Groundwaters*, Digest 250, Building Research Establishment, HMSO (June).

BRE (1989) *Brickwork: Prevention of Sulphate Attack*, Defect Action Sheet DAS 128 (April), Building Research Establishment.

BRE (1996) *Sulphate and Acid Resistance of Concrete in the Ground*, Digest 363, Building Research Establishment (January).

Neville, A.M. (1973) *Properties of Concrete*, 2nd edn, Pitman, London.

PSA (1989) *Defects in Buildings*, Property Services Agency, Department of the Environment, HMSO, p. 77.

Shirley, D.E. (1984) *Sulphate Resistance of Portland Cement Concrete: Measures to Prevent Sulphate Attack*, 2nd edn, Cement and Concrete Association.

CHAPTER 3

# Failure mechanisms in ceramic materials

# 3.1 Frost attack in bricks and concrete

## Learning objectives

- You should understand the mechanism of frost attack and the common symptoms and locations of failure in brickwork and concrete.
- It is also important to be able to identify the vulnerable areas of buildings.

Frost damage normally requires frequent freeze thaw cycles whilst the material is very wet. Hence it usually affects materials in direct or bridging contact with the ground, or subject to water run-off or collection. The degree of exposure of the detail is important, so parapets with both sides of their structure exposed represent a high risk, for instance.

The mechanism of frost attack is only partially dependent on the stresses induced by the hydrostatic pressure of water expanding as it freezes. However, very little correlation has been found between the saturation coefficient and the frost resistance of porous materials. (The saturation coefficient can be defined as the volume of pores filled in a saturation soaking test divided by the total pore volume.) Rather, the predominant factor appears to be pore diameter. This is thought to affect the rate of freezing of the pore water, with the larger pores freezing first whilst the unfrozen water in the smaller pores (and at a higher vapour pressure than ice) feeds the larger pores.

As this process continues, a build-up of hydrostatic pressure on the pore walls creates the tensile forces that damage the material. Hence it is the proportion of fine pores (and cul-de-sac pores) that is critical, and which the hard-firing of bricks minimizes (it is notable that underburning is also associated with high salt contents and hence problems with salt attack).

Frost attack occurs near the surface of materials, since moisture gains easy access through the open pore structure at the surface, and temperature fluctuations in the material are more dramatic. Hence temperatures which oscillate around freezing are more destructive than prolonged severe frosts.

*The Technology of Building Defects.* Dr John Hinks and Dr Geoff Cook.
Published in 1997 by E & FN Spon, 2–6 Boundary Row, London SE1 6HN, UK. ISBN 0 419 19770 2

**Fig. 3.1.** Severe weathering and frost attack of brickwork. (Reproduced from G. Cook and J. Hinks, *Appraising Building Defects*; published by Addison Wesley Longman, 1992.)

Frost attack in brickwork is usually progressive, causing delamination as the bricks become friable and fail in tension under the build-up of stresses. There appears to be little quantifiable link between compressive strength and frost resistance, except in extreme instances, since the frost attack operates through tensile stresses.

In conjunction with its friability, frost-attacked brickwork may exhibit expansion of the joints and possibly the bricks also. Longitudinal and vertical expansion of the walling will occur, leading to oversailing or bulging under restraint.

With concrete the ratio affects the pore size, and is therefore also important to its frost resistance. The significance of the ratio is non-linear, and increasing it above about 0.4 has a marked effect on durability.

## Revision notes

- Requires frequent freeze thaw cycles whilst material is very wet. Temperatures oscillating around freezing are more destructive than prolonged severe frosts.
- Material exposure is important; brick window cills and parapets are high risk, also direct or bridging contact with ground.
- Mechanism depends partly on hydrostatic pressure, primarily rate of freezing in larger pores, which small pores feed.
- Material fails in tension. Compressive strength bears little relationship to frost resistance.
- Failure produces delamination and friability. Oversailing and bulging may occur. Attack is progressive.

---

## Discussion topics

- Describe the mechanism of frost attack in porous materials, and produce an illustrated checklist of materials and details which are especially vulnerable to damage. Comment on any other causes of defects with which the symptoms of frost attack may be confused.

---

## Further reading

BRE (1987) *Concrete Part 1: Materials*, Digest 325, Building Research Establishment, HMSO (October).

Litvan, G.G. (1988) The mechanism of frost action in concrete: theory and practical implications. National Research Council, Canada. *Institute for Research in Construction Paper IRC 30420.*

Taylor, G.D. (1983) Frost resistance of porous building materials. *Building Technology and Management*, **21**(4), 30–31; **21**(5), 27–28; **21**(6), 8–9.

West, H.W.H. (1970) Clay products, in *Weathering and Performance of Building Materials* (eds J.W. Simpson and P.J. Horrobin), Medical and Technical Publishing, pp. 105–133.

# 3.2 Soluble salt crystallization

## Learning objectives

You should:

- understand the attack mechanism of crystallization in porous materials and the common symptoms of its occurrence;
- understand the methods of confirming the defect.

The crystallization of soluble salts is one of the biggest single problems affecting porous building materials, and especially brickwork. Stone and concrete are also vulnerable.

Aside from producing unsightly efflorescence on the surface of the material, the crystallization of some salts can cause extensive delamination of porous materials. The salts may originate from within the material itself, for example the sulphates contained in bricks; from a direct or bridging contact with the ground; or as pollutants in the surrounding atmosphere, for example road salt. Chlorides and sulphates produce particular problems which are discussed elsewhere in this book.

Efflorescence is fairly common with brickwork in the early stages of its life as the initial wetting/drying cycles leach out the free salts. Recurrent efflorescence in the external walls is usually indicative of repeated entry of water through faulty detailing, however.

The attack mechanism involves salt solutions passing through the pores of dry materials and leaving a deposition within the pores or on the surface upon evaporation. Capillary attraction may draw the salt solutions into narrow pores and allow their transportation to remote deposition sites. Hence a heavy deposition of salts may simply indicate a good drying area.

Salt deposition within the material gradually blocks the pores. Since the process requires cyclic wetting and drying, its rate is controlled by the exposure of the detail. The crystallizing salts accumulate at the depth of wetting and create a back pressure which delaminates the material in a manner similar to frost attack. Delamination exposes pores previously behind the wetting zone to progressive attack.

*The Technology of Building Defects.* Dr John Hinks and Dr Geoff Cook.
Published in 1997 by E & FN Spon, 2–6 Boundary Row, London SE1 6HN, UK. ISBN 0 419 19770 2

**Fig. 3.2.** The precipitation of soluble salts from brickwork at a DPC may even produce stalactite features.

Identification of the chemical composition may help identify the source, for instance nitrates or chlorides are usually indicative of external contamination.

## Revision notes

- Biggest single problem for porous building materials.
- Produces unsightly surface efflorescence. Severe attacks can create delamination.
- Sources of salts include the material under attack, groundwater and pollutants in the surrounding atmosphere.
- Early efflorescence is fairly common with brickwork. Recurrent efflorescence is usually indicative of faulty detailing.
- Salts deposited within pores accumulate at the depth of wetting and create a back pressure which delaminates the material. Delamination exposes pores previously behind the wetting zone to progressive attack.

## Discussion topics

- Compare and contrast the processes of delamination of porous materials by frost attack and deposition of soluble salts. Support your answer with sketches.

## Further reading

BRE (1974) *Clay Brickwork 2.* Digest 165, Building Research Establishment.

Higgins, D.D. (1982) Efflorescence on concrete, in *Appearance Matters*, Publication 47, Cement and Concrete Association, ch. 4.

Interpave (1995) *Efflorescence*, Precast Concrete Paving and Kerb Association.

MPA (1988) *Efflorescence and Bloom on Masonry*, Data Sheet No. 3, Mortar Producers Association.

Russell, P. (1983) *Efflorescence and the Discolouration of Concrete*, Viewpoint Publications

West, H.W.H. (1970) Clay products, *Weathering and Performance of Building Materials* (eds J.W. Simpson and P.J. Horrobin), Medical and Technical Publishing, pp. 105–133.

CHAPTER 4

# Failure mechanisms in timber

# 4.1 Defects in timber

## Learning objectives

You should be able to:

- describe timber as a natural fibre-reinforced material having variable properties;
- explain the differences between hardwoods and softwoods as being based on the natural features of the growing tree and not mechanical properties;
- describe a range of naturally occuring defects within the timber and explain their influence on the mechanical properties of timber;
- describe the anisotropic mechanical properties of timber;
- explain why timber may fail when incorrectly orientated to resist load;
- describe the agents of change which can adversly affect timber;
- discuss the influence of moisture content on the deterioration of timber.

## Problems with timber generally

Timber has been used extensively for construction purposes since the very earliest forms of timber shelter. It is still widely used in modern construction because of several useful properties, including ease of working, high strength-to-weight ratio and durability. The cell wall material of timber can have an elasticity similar to that of concrete, 27 000 $N/mm^2$, with a tensile strength of more than 140 $N/mm^2$. Greenheart, with a density of approximately 1040 $kg/m^2$, has been classified as 'very durable' with a life of 25 years, whereas balsa has a density of approximately 170 $kg/m^2$ and has been classified as 'perishable' with a life of less than 5 years.

These examples show that the properties of different timber vary considerably. There are, however several features which are common to many types of timber.

The strength of timber can be considerable, although it is important to identify which particular strength property is being described. The complex structure of timber has an influence on these strength differences. The structure can be generalized as a cylindrical bundle of straw-like fibrous

*The Technology of Building Defects.* Dr John Hinks and Dr Geoff Cook.
Published in 1997 by E & FN Spon, 2–6 Boundary Row, London SE1 6HN, UK. ISBN 0 419 19770 2

cells held together with band-like cells which radiate out from the centre. In the tree the straw-like cells run vertically up the trunk. The cell walls of these fibres contain significant amounts of cellulose, a polymer. Because of this structure, timber can be considered as a naturally occuring fibre-reinforced material.

The growth of the tree influences the density of the timber, with spring growth being less dense than summer growth. The growth area of timber, just below the bark in the cambium layer, lays down a ring of new sapwood. This layer acts as a conduit and storage facility for food and water. The inner heartwood layers of the tree offer structural support.

Timber is broadly classified into two types, softwoods and hardwoods. The straw-like cells in softwoods are termed 'tracheids' and these are passed through in the radial direction by ray cells at various intervals. The trees in general retain their leaves throughout the year and grow quickly; many can be felled within 30 years. Although the timber is generally of low density and durability, it is used extensively in construction. Treatment is required for softwood to be used externally; this may also reduce the risk of discolouration from ultraviolet radiation.

The hardwoods in general lose their leaves in winter, and to support the growth of new foliage in the spring vertical vessel cells act as food conduits. The straw-like cells are called fibres and are thicker-walled than tracheids. Rays pass through the fibres in a radial direction, although their structure and the general structure of hardwoods are more complex than that of softwood. Hardwoods can possess considerable durability.

### Note

There is a very large number of different types of timber and timber-based products; since a description of each is outside the scope of this book the reader is advised to consult the texts referred at the end of this section.

## Moisture content of timber

Moisture content (MC%) can be determined from:

$$\text{MC\%} = \frac{\text{Wet mass of sample} - \text{Dry mass of sample}}{\text{Dry mass of sample}} \times 100$$

where dry mass = dry mass of sample placed in an oven at 100–105°C until there is no further reduction in mass of the sample.

The moisture content of timber can be measured with meters which measure the electrical resistance of timber between two metal needles. These devices are reasonably accurate between 8% and 28% although the relationship between resistance and moisture content is likely to vary between timber species. Since the meters may be measuring the lowest resistance between the needles, there is always the risk of surface moisture influencing the results.

The growing tree may have a moisture content of 200%, although the sapwood is likely to have a highter moisture content than the heartwood. Moisture content is reduced through seasoning of the converted tree to levels applicable to the location and final use of the timber. This and other features of moisture content of timber are shown in Table 4.1.

**Table 4.1** Approximate moisture content of timber and physical effects

| Moisture content (%) | Description and physical effects |
|---|---|
| 28 | Fibre saturation point<br>>28%: no increase in size, timber in use is very wet indeed<br><28%: shrinkage occurs |
| 20 | Average level for external joinery<br>>20%: timber may be liable to fungal attack |
| 16 | Lower limit of natural seasoning<br>Average level of internal joinery in unheated interiors |
| 12 | Average level of internal joinery in heated interiors |
| 8 | Average level of internal joinery near heat emitters or in very hot interiors |
| 7 | Flooring over underfloor heating |
| 0 | Timber is oven-dry<br>Lower limit of shrinkage |

The fibres become saturated at a moisture content of around 28% and increases in moisture content above this value are not associated with volumetric expansion. The shrinkage which occurs below this value varies depending on the direction to be considered, the change in moisture content and the type of timber. It is suggested that a reduction in moisture content from 20% to 8% gives a 10% change in dimension tangentially, 6% radially and 0.1% longitudinally. These changes can cause distortion of the timber, which may set up internal stresses which the timber cannot resist and the material may split, check or shake. These changes can have detrimental effects, for example increases in the moisture content of timber floor coverings may cause the blocks to become detached from the sub-base. Floor blocks laid with a high moisture content may shrink when the building is in use.

Timber which is passing through wetting and drying cycles between widely spaced moisture contents is particularly vulnerable. The magnitude of creep of timber is influenced by the relative humidity of the surroundings, with high relative humidity and high loading causing high deflection in timber beams. The hygroscopic nature of timber means that the moisture content of the timber will be influenced by the relative humidty of the surroundings. This partially explains why timber in damp, unventilated spaces is at greater risk from fungal attack.

Moisture-movement-induced distortion is increased when there are differences in growth rates, slope of grain, directional moisture movement, rapid seasoning and poor stacking during seasoning.

Although the timber can be at risk from fungal attack at moisture contents of 20% it may continue below this level if the fungi can transport moisture, e.g. dry rot (*Serpula lacrymans*).

Where ferrous nails or truss plates are embedded into moist timber there is a risk of corrosion through electrolytic action. In addition, the hydroxyl ions produced on the surface of the nail can produce a localized alkaline environment, which can cause the timber to break down. This is because timber is generally acidic and has a reduced resistance to alkalis. Chemical attack of timber is possible, although this is commonly associated with flue gases. The sulphurous and nitrogenous gases cause hydrolysis of the timber surface, which separates the fibres.

## Mechanical and strength properties

Timber is anisotropic, since it has different mechanical properties depending on the direction being considered. Typical values are shown in Table 4.2.

**Table 4.2** Summary of typical mechanical properties of timber with a moisture content not more than 18%

| Timber type | Direction of loading in relation to grain direction | Type of loading | Strength ($N/mm^2$) |
|---|---|---|---|
| Douglas fir, imported (density 545 $kg/m^2$, MC 12%) | Parallel | Compressive | 53 |
| | | Tensile | 138 |
| | | Shear | 11 |
| | | Bending | 93 |
| | Perpendicular | Compressive | 6 |
| | | Tensile | 3 |
| Afromosia (density 72 l $kg/m^2$, MC 18%) | Parallel | Compressive | 22 |
| | | Tensile | 27 |
| | | Shear | 3 |
| | | Bending | 24 |
| | Perpendicular | Compressive | 6 |
| | | Tensile | 8 |

It is essential to use timber in the correct orientation. Where structural failure of a timber has occured the orientation and loading directions should be determined.

Since published strength properties are based on small specimens of timber having no obvious growth defect the actual strength of real timber in a construction application is likely to be less.

### Note

A detailed appraisal of the mechanical strength of timber is outside the scope of this book. Readers are advised to consult the references shown at the end of this section.

There is a need to review some of the key defects which affect the mechanical performance of timber, including the following.

## The influence of rate of growth

Assessed by counting the growth rings per centimetre. High numbers indicate slow growth and higher-density timber than samples with a low number of growth rings. The typical minimum for softwood is 4 growth rings per cm. It is suggested that the compressive strength of clear timber parallel to the grain is 100 times its relative density. Hardwoods in general show high rates of growth in the spring wood and slower growth in the summer wood.

## Moisture content

Lowering the moisture content of timber results in an increase in strength. This increase varies, depending on direction and type of timber. Typically there is a 40% increase in strength when the moisture content is reduced from fibre saturation point to 15%, and a 300% increase when reduced to 0%. The assessment of moisture content has structural as well as durability implications.

## Slope of grain

Since a typical tensile strength of a softwood perpendicular to the grain can be 3 $N/mm^2$, and parallel to the grain 138 $N/mm^2$, it is clear that the slope of the grain has a considerable influence on structural performance. Even a 10° slope of grain in timber can cause a significant reduction in bonding strength.

**Fig. 4.1.** External structural timber cracking along the grain. (University of Reading.)

## Problems with knots

These are either living or dead branches from the growing tree. Dead knots may fall out of thin timber sections, emphasizing their lack of contribution to the strength of the timber. In areas of low stress, e.g. in the centre of beams and joists, this may not be a problem. In areas of tensile stress, owing to the localized variation in slope of grain as it passes around the knot, their location is more problematic. Visual assessment of knots is complicated by their three-dimensional nature, and the need to inspect all faces of the timber section.

## Fire

Although timber burns, the charring of the external surfaces acts as an insulator and moisture in the timber must be evaporated before burning can progress. This slows down the rate of burning, and for large-sized timbers may mean that significant amounts of mechanically sound timber remain in the centre of the section. This complicates the visual assessment of the structural integrity of timber sections which have been subject to fire.

### Revision notes

- Timber is a naturally occurring fibre-reinforced material and is likely to have variable properties, even within a specimen of timber.
- The differences between hardwoods and softwoods are based on the natural features of the growing tree and not mechanical properties.
- There are many agents of change which can affect timber, adversely affecting the mechanical properties. These include chemical attack, weathering, fungi and insects.
- Since timber has anisotropic mechanical properties failure may occur when the timber is incorrectly orientated to resist load.
- In use the influence of moisture is critical since it affects dimensional stability, resistance to fungal attack, creep and strength.

---

## ■ Discussion topics

- Compare the deterioration mechanisms which may affect hardwood and softwood.
- Discuss the implications for the deterioration of softwood timber of changes in moisture content.
- Describe how the mechanical properties of timber may contribute to failure of a softwood floor joist.

- Compare the mechanical failure mechanisms associated with fibre-reinforced materials with those of timber.

---

## Further reading


BRE (1974) *Drying Out Buildings*, Digest 163, Building Research Establishment, HMSO.

BRE (1985) *Timbers: Their Natural Durability and Resistance to Preservative Treatment*, Digest 296, Building Research Establishment, HMSO.

BRE (1990) *Painting Exterior Wood*, Digest 354, Building Research Establishment, HMSO.

Cook, G.K. and Hinks, A.J. (1992) *Appraising Building Defects: Perspectives on Stability and Hygrothermal Performance*, Longman Scientific & Technical, London.

Desch, H.E. (1981) *Timber: Its Structure, Properties and Utilisation*, 6th edn (revised by J.M. Dinwoodie) Macmillan Education, London.

Everett, A. (1975) *Materials – Mitchells Building Construction*, Longman Scientific & Technical, London.

Princes Risborough Laboratory (1973) The Natural Durability Classification of Timber. *Technical Note No. 40*, HMSO.

PSA (1989) *Defects in Buildings*, HMSO.

Ransom, W.H. (1989) *Building Failures Diagnosis and Avoidance*, 2nd edn, E. & F.N. Spon, London.

Richardson, B.A. (1991) *Defects and Deterioration in Buildings*, E. & F.N. Spon, London.

Taylor, G.D. (1991) *Construction Materials*, Longman Scientific & Technical, London.

# 4.2 Fungal attack of timber

## Learning objectives

- You should be aware of the broad categories of fungal attack and the necessary conditions for attack to commence and be sustained.
- The difficulty in correctly identifying rots should be appreciated.

Fungi usually attack untreated wood by developing extensive root systems (mycelium) to remove nutrients and oxygen from the cells, so destroying the chemical structure of the host timber. As the roots extend throughout the cellular structure, the timber disintegrates and becomes more susceptible to secondary damage such as insect attack. The presence of fungi may be indicative of a more critical dampness problem.

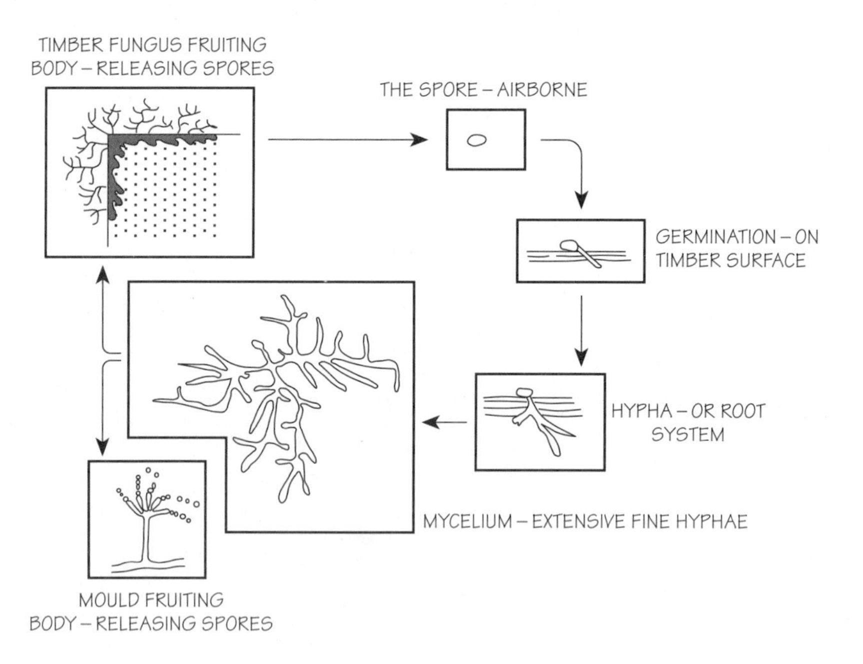


**Fig. 4.2.** The growth cycle of timber fungi.

*The Technology of Building Defects.* Dr John Hinks and Dr Geoff Cook.
Published in 1997 by E & FN Spon, 2–6 Boundary Row, London SE1 6HN, UK. ISBN 0 419 19770 2

The risk of attack depends largely on the species of timber and the proportion of sapwood, and also on the surroundings and the type of fungal spores present. Chronic dampness occurring in conjunction with poor ventilation and a normal internal temperature range represent ideal conditions for potential fungal attack. Not surprisingly, many of the locations prone to rot are also the most difficult to survey.

**Fig. 4.3.** Dampness and fungi can also affect building services. (S.D. McGlynn.)

Wood damaged by fungus may sound hollow and dull, and be of abnormal colour. It will yield easily under the pressure of a sharp instrument, tending to break across the grain.

Wood-rotting fungi are generally categorized as wet or dry rots. Both dry and wet rots will attack hardwood and softwoods and, whilst dry rots will not grow in saturated conditions, neither will grow in a dry environment. The colour and scale of the fruiting body and spores can be used to differentiate fungi.

Dry rot (*Serpula lacrymans*) is a brown, cubical rot that attacks the cellulose and lignin in wood, converting it into a brownish-red dry powder. In damp atmospheres with poor ventilation the rot develops a fluffy, cotton-wool appearance. Advanced dry rot is accompanied by a musty or mushroomy smell.

Dry rot is widely considered as the most serious of the wood rots, because of its structural significance and pervasiveness. Remedial works can be extensive and not limited to the brickwork. In dry conditions the surface appearance of the rot can be greyish, with bright yellow mycelium where there is light. Where the timber supply has been exhausted only the hollow strands used for moisture transport remain. These are called hyphae, and vary in size from 1 mm to 10 mm in diameter. They may cross or penetrate non-cellulosic materials in their path, leaving sites of dormant fungus.

Fruiting bodies (sporophores) develop in the final stages of the life cycle and are identifiable by their white perimeter. The surrounding timber may be coated in a fine dusting of red spores which are easily dislodged from the

**Fig. 4.4.** General view of the failure, due to dry rot, of a suspended timber ground floor. (S.D. McGlynn.)

fruiting body. In still air the fruiting bodies may exude water, and it is from this characteristic that the name *lacrymans* (crying) is derived.

The wet rots are less pervasive and generally less harmful than dry rot, but can become widespread in excessively damp conditions. In general a minimum moisture content of 20% is necessary for germination of their spores. A continuous supply of moisture is needed for the transport of the fungal enzymes into the cellulosic timber and the dissolution of the resulting glucose for return to the fungus. Hence moisture contents below 18–20% are widely considered to be secure. Extreme dampness excludes oxygen from the wood and limits the rots, as do temperatures above about 40°C. Fluctuations in moisture content may drive the fungi into dormancy.

The most frequently occurring wet rot is *Coniophora puteana* (formerly *Coniophora cerebella*), also known as cellar fungus. It produces a cubical

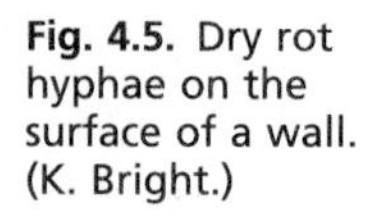

**Fig. 4.5.** Dry rot hyphae on the surface of a wall. (K. Bright.)

failure in the wood that is difficult to distinguish from the superficial symptoms of dry rot, although the fruiting body may be characteristically olive green or brown. The wet rot darkens the affected timber, which also becomes very brittle and easily powdered. It may crack parallel to the grain. Frequently the deeper, damper wood decays whilst a skin of deceptively sound timber remains.

*Phellinus continguus* (formerly known as *Poria vaillantii*) is actually categorized as a dry rot and has a very similar appearance to that of *Serpula lacrymans*, but requires very damp conditions and is only likely to produce localized attack in saturated conditions. A key distinction is that in the dry state the fruiting bodies remain leathery (instead of becoming brittle as does *Serpula lacrymans*).

The hyphae associated with sap stain may present themselves as a deeply penetrating blue or bluish-grey streaking. Although the immediate structural relevance of sap stain may be limited, the attack usually occurs in sap-rich wood (the type which is most susceptible to fungal attack) and may be a precursor to and early warning of other fungal attack.

## Revision notes

- Timber disintegrates as extensive fungal root systems remove the nutrients and oxygen from the cells.
- Risk of attack depends on species of timber, proportion of sapwood, surroundings and type of fungal spores.
- Coincident chronic dampness, poor ventilation and normal internal temperatures produce ideal conditions for fungi.
- Wood disintegrates, may become abnormally coloured and sounds hollow and dull. Fruiting body colours aid the identification of fungus.
- Dry and wet rots will attack hardwood and softwoods and, whilst dry rots will not grow in saturated conditions, neither will grow in a dry environment. Possibility of dormancy.
- Dry rot: brown, cubical, fluffy, with yellow mycelium, red spores and a musty smell. Fruiting body may weep. Can be structurally serious. Hyphae will cross brickwork.
- Wet rots: generally less harmful than dry rot, need chronic damp conditions. Safe below moisture contents of 18–20% and temperatures above 40°C.
- Cellar fungus produces cubical failure in the wood similar to dry rot, olive green or brown fruiting body. Affected timber darkened, brittle. Surface may appear to be sound.

## ■ Discussion topics

- Produce a tabular comparison of the conditions required for the various forms of fungal attack of timber, the symptoms of such attacks, the likely locations and suggestions for remedial work.
- Compare and contrast the forms of attack and consequences of dry and wet rots in constructional timber. Describe how the timber supports the various stages of the fungal life cycle.

## Further reading


Addleson, L. (1992) *Building Failures: a Guide to Diagnosis, Remedy and Prevention*, 3rd edn, Butterworth Architecture, Study 27: Timber decay, pp. 163–164.

Berry, R.W. (1994) *Remedial Treatment of Wood Rot and Insect Attack in Buildings*, Building Research Establishment.

Bravery, A.F. (1985) *Mould and its Control*, BRE Information Paper IP 11/85.

Bravery, A.F., *et al.* (1992) Recognising wood rot and insect damage in buildings. *BRE Report 232*.

BRE (1977) *Decay in Buildings: Recognition, Prevention and Cure*, Technical Note 44, Building Research Establishment.

BRE (1988) *Home Inspection for Dampness. First Step to Remedial Treatment for Wood Rot*, BRE Information Paper IP 19/88.

BRE (1989) *Dry Rot: Its Recognition and Control*, Digest 299 Building Research Establishment.

BRE (1989) *Wet Rots: Recognition and Control*, Digest 345, Building Research Establishment (June).

BRE (1990) *Surface Condensation and Mould Growth in Traditionally Built Dwellings*, Digest 297, Building Research Establishment.

BRE (1991) *Timber Decay and Control*, Technical Note 53, Building Research Establishment.

Coggins, C.R. (1989) *Decay of Timber in Buildings*, Rentokil Ltd.

Gibson, A.P. and Lothian, M.T. (1982) Surveying for timber decay. *Structural Survey* **1**(3), 262–268 (Autumn).

Hollis, M. and Gibson, C. (1990) *Surveying Buildings*, RICS Books, pp. 211–228.

Lea, R.G. (1992) *Wood-based Panel Products: Moisture Effects and Assessing the Risk of Decay*, BRE Information Paper IP 19/92.

Richardson, B.A. (1991) *Defects and Deterioration in Buildings*, E. & F.N. Spon, London, pp. 84–89.

TRADA (1984) *Timber Pests and Their Control*, Timber Research and Development Association, British Wood Preserving Association.

# 4.3 Insect attack of timber

## Learning objectives

- You should be aware of the broad categories of insect attack and the necessary conditions for attack to commence and be sustained.
- The difficulty in correctly identifying insect attacks should be appreciated.

Alongside fungal attack, insects are one of the biggest causes of timber decay. The main external symptoms of insect attack are the flight holes formed by adult insects leaving the wood. These vary in size and spacing, and can therefore be used as an indicator of the insect responsible. However, flight holes may be located on the concealed faces of timber (especially with floorboards), so some disassembly may be involved in an inspection.

The main structural damage to the wood is internal, caused by the insect larvae extracting the glucose from the cells, leaving frass and wormholes. Timber with a high sapwood content is a frequent target.

Whether the attack is continuing is obviously an important but difficult detail to establish, and a full inspection of the building will be necessary to ensure the identification of all affected timbers. The scattering of bore dust around flight holes may indicate an active infestation.

The common beetles causing damage in this country are the house longhorn beetle, the powder post beetle, the common furniture beetle and the death-watch beetle. Wood-boring weevils and bark borers may also be encountered.

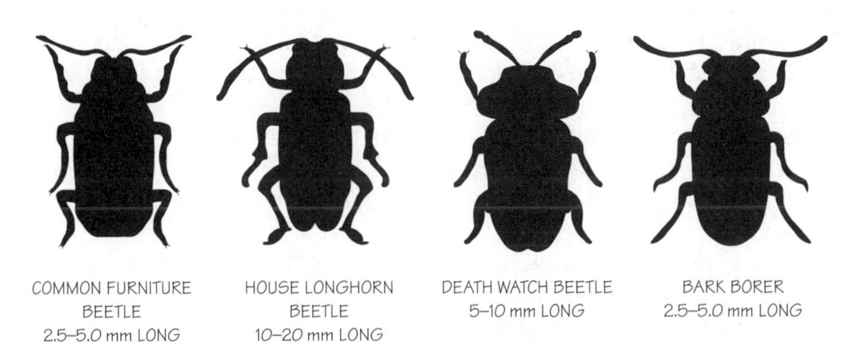


**Fig. 4.6.** Insects which attack timber.

*The Technology of Building Defects.* Dr John Hinks and Dr Geoff Cook.
Published in 1997 by E & FN Spon, 2–6 Boundary Row, London SE1 6HN, UK. ISBN 0 419 19770 2

The house longhorn beetle (*Hylotropes bajulus*) usually only thrives in tropical conditions, but mating and infestation in temperate climates are occasional and they may afflict timber in place or under storage. Attack is usually limited to the sapwood in softwoods, and occurs under the surface of the timber.

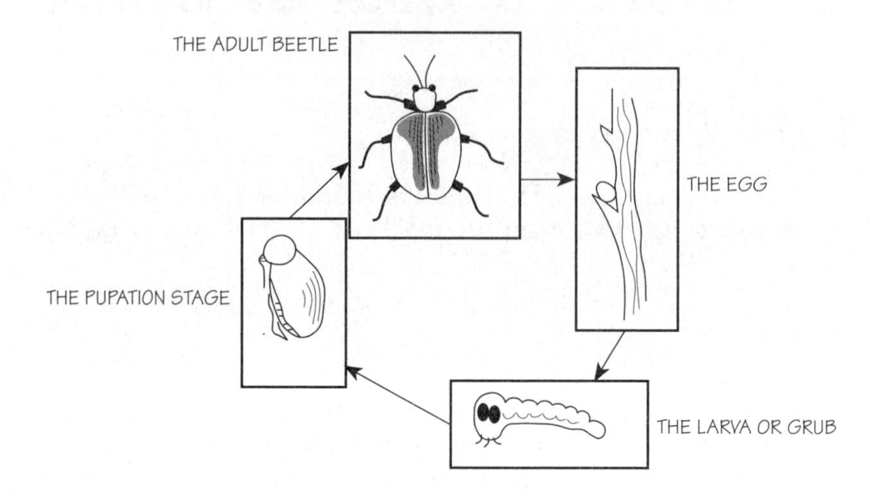


**Fig. 4.7.** Life cycle of timber insect.

The life cycle is long, commonly between five and seven years, so it is quite possible for infested timber to be apparently sound. Established attack may produce some surface bulging, and may be structurally serious enough to require replacement of members, in floors and particularly in roof spaces. It is important that all infected timber is cut back thoroughly.

The beetle leaves oval flight holes in the surface of the timber, which are relatively large (3–9 mm diameter). Adult beetles may be 20 mm long.

There are several species of powder post beetle (*Lyctus*), which attack the sapwood of starchy hardwoods such as elm, oak, ash and chestnut. Attack may progress under an apparently sound surface, and tends to occur in sapwood-rich timber.

*Lyctus* has a life cycle of between one and two years. It usually emerges in the spring and summer months, leaving a complete destruction below an otherwise intact surface veneer. The insect is also characterized by its particularly fine frass.

The common furniture beetle (*Anobium punctatum*), also known as woodworm, is probably responsible for most of the insect damage to timber in the UK. Infestation is thought to be profuse and is characterized by a gritty, cream-coloured, cigar-shaped frass. The life cycle can vary between two and five years, with small circular flight holes (1.5 mm) appearing during the summer. The damage is produced by the burrowing larvae and may be progressive.

Aside from furniture, hardwood and cheap softwood carcassing timbers (with a high sapwood content) may be vulnerable. The furniture beetle seems to prefer chronically damp timber, and very dry wood seems to be less susceptible.

The death-watch beetle (*Xestobium rufovillosum*) frequently attacks hardwoods such as oak, chestnut and elm, and may be found in older timber

**Fig. 4.8.** This tie connection no longer exists, owing to a sustained insect attack.

structures. With common life cycles of 5–6 year duration and the larvae showing resilience to dry conditions, infestation is widespread throughout the warmer parts of UK.

Attack is common in damp, concealed and poorly ventilated timbers, such as those susceptible to rot (the existence of which appears also to encourage attack). Hence infestation may commence in concealed sappy roof timbers or wall plates exposed to leakages, and quickly spread to other timber. Attack is usually structurally significant and a thorough inspection of structural elements is essential. Flight holes are of moderate size (2–3 mm diameter), and the timber contains oval discs of frass.

Wood-boring weevils (*Pentarthrum huttoni*, *Euophyrum confine*) also favour wood suffering from damp and/or fungal attack. Hence common locations for infestation include below-ground timbers such as floor joists and timbers in contact with damp walls, such as wall plates, masonry and built-in joists. Plywood may also be attacked.

The insect totally destroys the integrity of the timber and leaves a small flight hole (less than 1 mm diameter).

Bark borers (*Ernobius mollis*) infest the sapwood of softwoods. Attack is commonly located in roof timbers and floor joists, but usually creates no structural damage. It is therefore important to distinguish it from destructive insects such as furniture beetle which produce similar symptoms.

## Revision notes

- Main external symptoms are flight holes, which vary in size with insect. Structural damage is internal, usually to sapwood. Residues of frass and dust.

- Important to establish whether attack is ongoing.
- **House longhorn**: softwood sapwood attack occasional in Home Counties, roof and floor timbers common locations, adults up to 20 mm, five-to-seven-year life cycle, large oval flight holes. Activity most likely between June and August.
- **Powder post**: attacks the starchy sapwood edges of hardwood, surface may appear sound over complete destruction. One-to-two-year life cycle.
- **Common furniture** (woodworm): profuse and preferring damp or fungally infected sites. Progressive attacks on sapwood of softwoods mostly, although hardwoods may also be attacked. Activity most likely between June and August. Infestation is characterized by a gritty, cream-coloured, cigar-shaped frass.
- **Death-watch beetle**: frequently attacks hardwoods such as oak, chestnut and elm. May be found in older timber structures. Moderate flight hole, activity is most likely between April and June. Usually attacks areas exposed to leakage or in contact with damp masonry. Usually structurally significant. Attack may be preceded by fungal attack.
- **Wood-boring weevils**: frequently attack timber already suffering from damp and/or fungal attack. Total destruction of integrity and strength. Small flight hole, and insect is characterized by a prominent snout.
- **Bark borers**: attack is commonly located in roof timbers and floor joists, usually creates no structural damage.

## Discussion topics

- Produce a tabular comparison of the conditions required for the various insects which attack timber, noting the symptoms of such attacks, the likely locations and the vulnerable timbers.
- Compare and contrast the forms of attack and consequences of insect attacks in constructional timber. To what extent is it possible to correlate the age, nature of construction and type of location/ components of the building with the nature of attack?

## Further reading

Berry, R.W. (1994) *Remedial Treatment of Wood Rot and Insect Attack in Buildings*, Building Research Establishment.

Bravery, A.F. *et al.* (1992) Recognising wood rot and insect damage in buildings. *BRE Report 232.*

BRE (1970) *The Death Watch Beetle*, Technical Note 45, Building Research Establishment.

BRE (1972) *Damage by Ambrosia (Pinhole Borer) Beetles*, Technical Note 55, Building Research Establishment.

BRE (1977) *The Common Furniture Beetle*, Technical Note 47, Building Research Establishment.
BRE (1977) *The House Longhorn Beetle*, Technical Note 39, Building Research Establishment.
BRE (1992) *Identifying Damage by Wood-boring Insects*, Digest 307, Building Research Establishment.
BRE (1992) *Recognising wood rot and insect damage in timber*, Report BR232, Building Research Establishment.
Coggins, C.R. (1980) *Decay of Timber in Buildings*, Rentokil Ltd.
Desch, H.E. and Desch, S. (1970) *Structural Surveying*, Griffin.
Gibson, A.P. and Lothian, M.T. (1982) Surveying for timber decay. *Structural Survey* **1**(3) 262–268.
Hickin, N.E. (1975) *The Insect Factor in Wood Decay*, Associated Business Programmes Ltd.
Hollis, M. and Gibson, C. (1990) *Surveying Buildings*, RICS Books, pp. 195–210.
Lea, R.G. (1982) *House Longhorn Beetle Survey*, BRE Information Paper IP 12/82 (July).
Lea, G.R. (1994) *House Longhorn Beetle – Geographical Distribution and Pest Status in UK*, BRE Information Paper IP 8/94.
Lea, G.R. (1995) *Cockroach Infestations of Dwellings in U.K.*, BRE Information Paper IP 1/95.
Mika, S.L.J. and Desch, S.C. (1988) *Structural Surveying*, 2nd edn Macmillan Education.
Read, S.J. (1986) *Controlling Death Watch Beetle*, BRE Information Paper IP 19/86.
Richardson, B.A. (1991) *Defects and Deterioration in Buildings*, E. & F.N. Spon, London, pp. 84–89.
TRADA (1984) *Timber Pests and Their Control*, Timber Research and Development Association, British Wood Preserving Association.
White, M.G. (1970) *The Inspection and Treatment of Houses for Damage by Wood-boring Insects*, TimberLab Paper 33.

CHAPTER 5

# Failure mechanisms in metals

## Learning objectives

You should be able to distinguish between:

- the various mechanisms of corrosion and the factors involved: electrolytic corrosion, sacrificial protection and pitting corrosion,
- the symptoms of corrosion in exposed and concealed metals used for building.

Aside from cost and design reasons, the specification of metals as external claddings to buildings is usually made in recognition of their durability. However, most metals are not without significant potential deficiencies, realized when oxidation or reaction with other chemicals occurs.

Aggressive environments inside or outside the building can lead, for instance, to the deposition of sulphurous acids on claddings and roofs. Industrial processes conducted inside buildings can deposit a further range of aggressive compounds onto metals. When in aqueous solution these deposits may form the electrolyte required for the establishment of corrosion cells. Note that the moisture required to initiate such processes may be present as condensation or even drawn out of the air by some hygroscopic compounds.

The vulnerability of metals to alkali environments is not consistent, however. Steel encased in cementitious materials containing lime will be preserved by the highly alkaline environment. In contrast, lead and aluminium are etched by alkaline water running off from cementitious materials, and where concrete structures support metal sheets and components, a risk of condensation corrosion exists. This risk is enhanced where vapour barriers are omitted, inadequately positioned or jointed.

The occurrence of reversible movement in metals is directly related to their coefficient of expansion. This can be significant, since typical values are $17.1 \times 10^{-6}$ per °C for copper and $29.7 \times 10^{-6}$ for lead. Failure to allow for this at joints and changes of direction will produce buckling, distortion and cracking.

In the case of metal sheeting, any flexure of the background imposes additional bending stresses which can also cause deformations or cracking.

Decking and supporting materials which have different moisture and/or temperature movement coefficients will produce differential movement stresses at the interface, which may lead to forced dimensional change. With insulated deckings, the exposed metal surfaces can suffer extreme temperature variations (and movement stresses) caused by solar gain. The interface becomes a key movement zone.

*The Technology of Building Defects.* Dr John Hinks and Dr Geoff Cook.
Published in 1997 by E & FN Spon, 2–6 Boundary Row, London SE1 6HN, UK. ISBN 0 419 19770 2

The long-term loading of metals may produce additional straining especially in large panels where the fixing and support may be sparse. Such behaviour is termed 'creep', and may show itself at joints and fixing points. Soft metals such as lead may exhibit creep under their self-weight according to their elasticity and further test the large tolerances required of their fixing details. Laps and vertical drips in metal roofs are particularly vulnerable to creep damage.

Metals subjected to cyclic loading and unloading can suffer from fatigue, transforming the highly stressed zones of a ductile metal component into brittle weak spots. These loads may arise from self-weight, thermal movement and live or imposed loads.

## Corrosion of steel

Unprotected steel will corrode rapidly in wet and oxygenated conditions, the rust forming as a result of a complex set of reactions culminating in the production of hydrated ferric oxide ($Fe_2O_3 \cdot H_2O$). Hence steel piles, with minimal oxygen levels in the subsoil surroundings, corrode very slowly regardless of the nature of the soil. Obviously, the corrosion rate will be different above ground and at the interface.

The corrosion of reinforcement steel operates through an electro chemical process determined in part by the presence of differential oxygen levels within the reinforced concrete. Different regions act as cathodes or anodes according to whether oxygen is present or not.

The occurrence of differential oxygen concentrations in pore water and anionic solutions of pollutants such as carbonates, chlorides or sulphates

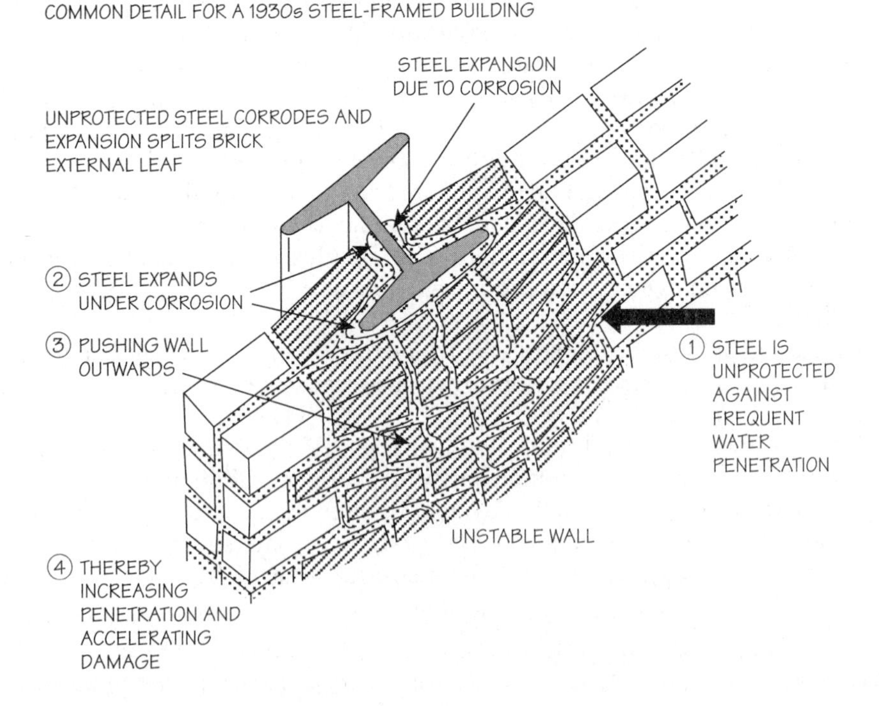


**Fig. 5.1.** Corrosion in unprotected mild steel.

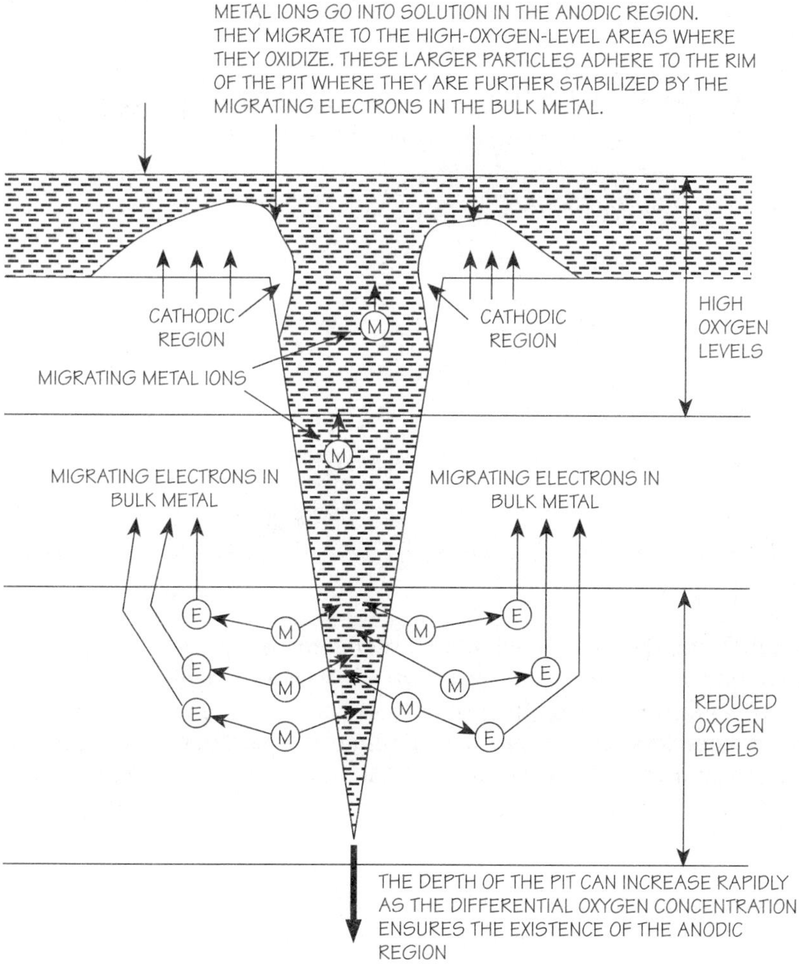

**Fig. 5.2.** The pitting corrosion of mild steel.

leads to the creation of microgalvanic corrosion cells. As these corrosion cells destroy the thin protective oxide film formed on the surface of steel in alkaline environments such as concrete, so corrosion takes place. Clearly the occurrence of carbonation, chloride and/or sulphate attack increases the likelihood and severity of corrosion in steelwork. Other factors affecting the corrosion resistance of the steel include the effective depth of cover to the reinforcement and the porosity of the concrete.

Note that in contrast to steel, the oxides formed from some metal ions occupy less space than the original metal. This produces a porous oxide layer which allows oxidation to continue into the body of the metal.

Galvanic cells can also be established between different metals (e.g. steel and lead) which are in contact directly or via a conductive medium. For instance, contact between reinforcement steel and silicon bronze cladding fixings leads to corrosion of the steel as it behaves as a sacrificial (oxidizing) anode.

Aside from the loss of tensile strength, the key consequences of corrosion in reinforcement steel are the accompanying disruptive expansion in the concrete and loss of bond strength between the steel and concrete. Hence

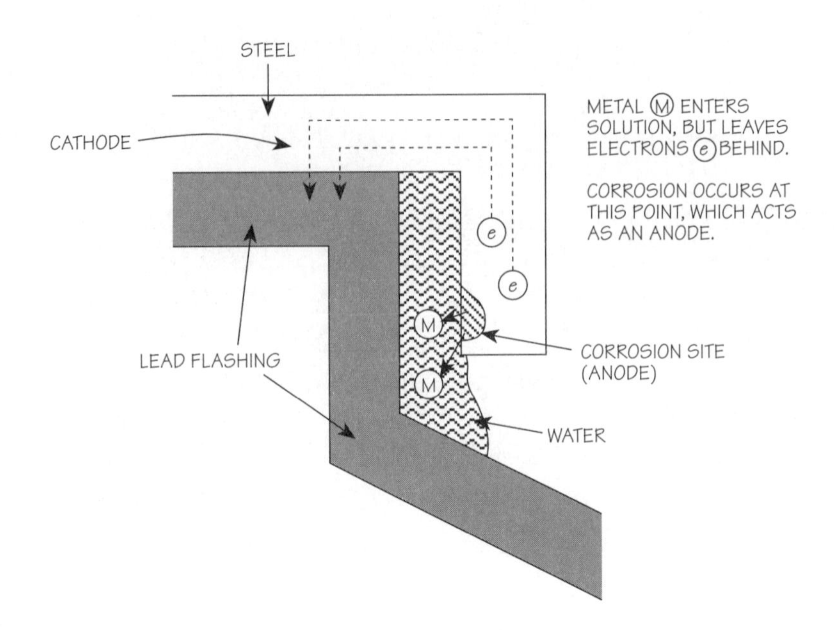


**Fig. 5.3.** Electrolytic corrosion.

corrosion in the tendons of post-tensioned components can be very serious, especially if the grouting is damaged by corrosion expansion.

An early symptom of reinforcement corrosion is the appearance of thin, directional cracks following the line of reinforcement, perhaps accompanied by pattern staining at the surface. Note that in instances where the iron staining at the surface is irregular, then iron content in the aggregates may be responsible.

The thin cracks develop into the displacement of sections of the covering concrete, culminating in spalling. Since the cracking leads to increased water penetration, the problem is self-perpetuating.

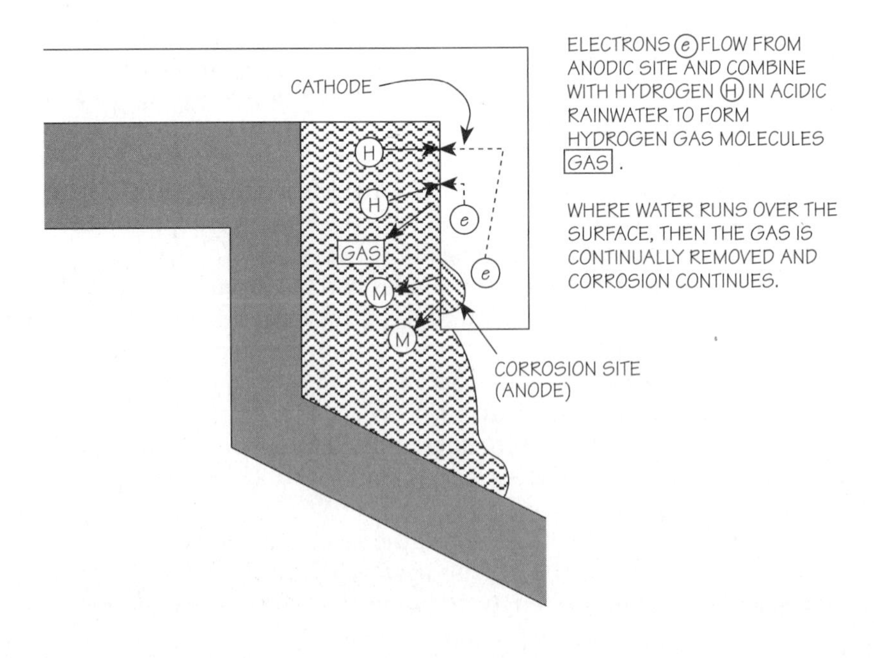


**Fig. 5.4.** Acidic corrosion.

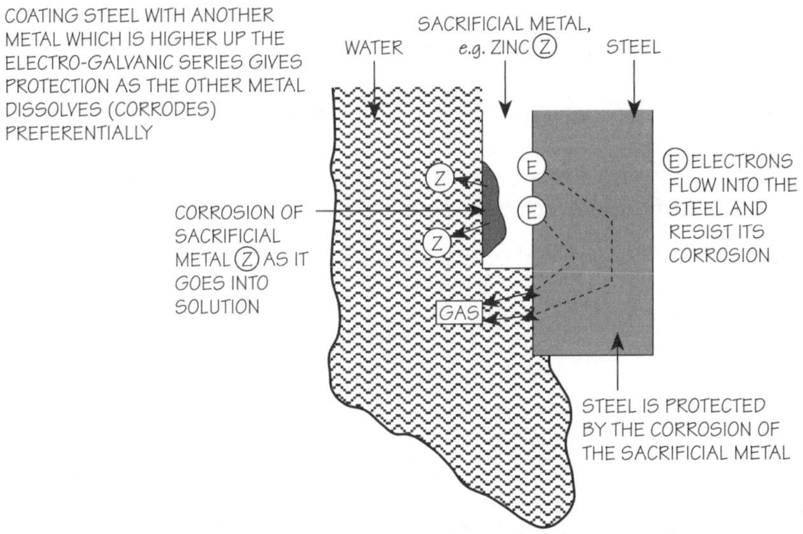

**Fig. 5.5.** Sacrificial protection.

Exposed steel can be protected by galvanizing. Here a surface layer of zinc corrodes preferentially to electrochemically protect the steel for the period it takes to dissolve the zinc. The thicker the zinc, the more durable the steelwork. Any perforations in the coating or variable thickness will affect the durability locally.

On exposed steelwork the detailing can make a significant difference to the durability. Hence stiffeners which create catchment areas for water are problematic. Environmental conditions are also a key feature in the occurrence and rate of corrosion.

Turning to exposed metals generally, the issue of patination arises. The pale green surface appearance of patinated copper is generally interpreted as enhancing buildings. The patina is an oxidized surface coating, and is a term which originates with the green encrustation forming on bronze, but which occurs with other metals including stainless steel, zinc, lead and

**Fig. 5.6.** Corrosion of external metalwork. Joints and other stressed regions are particularly vulnerable. (University of Reading.)

aluminium. Where water run-off occurs the patina may be washed to re-deposit on adjacent surfaces. Clearly, damage by erosion will occur to the patinating metal.

Localized corrosion may produce surface pitting leading to rapid penetration of the metal. This type of attack in steelwork may indicate chloride attack. Aluminium is also vulnerable to localized pitting caused by attacks involving environmental pollution.

## Revision notes

- Aggressive chemical deposits on metals may enter solution and form electrolytic corrosion cells.
- Moisture required for such processes may occur as direct water run-off, condensation or by abstraction from the air.
- Vulnerability to attack by acidic and/or alkalined environments is not consistent and depends on the metal involved. Steel is protected by alkaline environment, lead and aluminium are attacked.
- Metals may also be attacked by organic compounds such as the acids in certain timbers (which can attack zinc) and organic compounds in some bitumen products (which can attack copper).
- Failure to allow for significant reversible movement in metals at joints and changes of direction will produce buckling, distortion and cracking.
- Background flexure imposes additional bending stresses causing deformations or cracking.
- Decking and supporting materials with different moisture and/or temperature movement coefficients will produce differential movement stresses at the interface.
- With insulated deckings, the exposed metal surfaces can suffer extreme temperature variations (and hence movement stresses) caused by solar gain.
- Metals subjected to cyclic loading and unloading can suffer from fatigue, termed 'work-hardening'.
- Creep is a time-dependent strain produced by long-term loads. Damage is likely at joints and fixing points. Alloys may have increased resistance to creep.
- A patina is an oxidized surface coating. Patination occurs on copper and other metals including stainless steel, zinc and aluminium; the patination colour of the latter acts as a pollution indicator.
- Localized corrosion may produce surface pitting.

### Corrosion of steel

- Corrosion dependent on the presence of oxygen and damp. Electrochemical corrosion cells may be formed where differential oxygenation levels occur.

- Pollutants such as carbonates, chlorides or sulphates also lead to the creation of microgalvanic corrosion cells.
- Galvanic cells can also be established between different metals which are in contact directly or via a conductive medium.
- Corrosion in reinforcement steel depends on depth of cover, quality of concrete and presence of pollutants.
- Consequences include loss of tensile strength, disruptive expansion and loss of bond strength between steel and concrete.
- Early symptoms include thin directional cracking, followed by spalling.

## Discussion topics

- Compare and contrast the various mechanisms for corrosion in the metals in widespread use for construction.
- Describe the process of corrosion in steel, distinguishing between the mechanisms for exposed and encased steelwork. On the basis of this, produce a checklist for good practice in construction involving steel.
- Compare and contrast the various mechanisms of potential failure occurring in the metals in widespread use for construction. On the basis of this, produce an idealized checklist for the inspection of the metallic components of typical domestic buildings from several different construction eras and comment on the significance of the limitations in its use.

## Further reading

Bonshor, R.B. and Bonshor, L.L. (1996) *Cracking in Buildings*, Construction Research Communications (BRE), pp. 10, 59–63.

BRE (1982) *The Durability of Steel in Concrete: Part 1, Mechanism of Protection and Corrosion*, Digest 263 (July), Building Research Establishment.

BRE (1982) *Durability of Steel in Concrete: Part 2: Diagnosis and Assessment of Corrosion Cracked Concrete*, Digest 264 (August), Building Research Establishment.

BRE (1983) *Cracking and Corrosion of Reinforcement*, Digest 389 (December), Building Research Establishment.

Butlin, R.N. (1989) *The Effects of Acid Deposition on Buildings and Building Materials in the United Kingdom*, Building Research Establishment, HMSO.

Corrosion of steel in concrete (1988) *Report of the Technical Committee 60 CSC RI – Hove*. Chapman & Hall, London.

de Vekey, R.C. (1990) *Corrosion of Steel Wall Ties: History of Occurrence, Background and Treatment*, BRE Information Paper 1P, 12/90.

de Vekey, R.C. (1990) *Corrosion of Steel Wall Ties: Recognition and Inspection*, BRE Information Paper 1P, 13/90.

Hollis, M. and Gibson, C. (1990) *Surveying Buildings*, RICS Books, pp. 143–146.

Hudson, R.M. (1988) The effect of environmental conditions on the performance of steel in buildings. *Structural Survey* **6**(3), 215–223.

Society of Chemical Industry (1988) *Corrosion of Reinforcement in Concrete Construction*, Harwood, Chichester.

Taylor, G.D. (1983) *Materials of Construction*, Construction Press.

PART 2

# Components

CHAPTER 6

# Defects in components: general mechanisms

# 6.1 Moisture movement

## Learning objectives

- You should be able to explain the cause and possible consequences of moisture movement, distinguishing between reversible and irreversible moisture movement phenomena.
- The possible interaction of moisture induced movement with other causes of movement should be understood also.

Moisture movement is a natural and common phenomenon affecting building components. It is also one of the most fundamental and major sources of defects in building components and elements.

Moisture movement can occur as a discrete problem or in conjunction with other causes of movement, such as thermal movement, to produce combinations of symptoms.

In general, moisture movement is a phenomenon affecting traditional, permeable materials. Impermeable materials do not exhibit moisture movement (but may suffer other moisture-related problems connected with the shedding or penetration of rainwater at interfaces).

The variability in the permeability of traditional building materials produces a wide range of possible moisture movements. In the extreme, very permeable materials tend to respond quickly and dramatically to changes in moisture content. Their associated moisture movements may cause particular problems at the connections between components or elements.

The basic mechanism of moisture movement in materials and components is the swelling or shrinkage of a material as it harmonizes its moisture content with the ambient moisture level. This form of moisture movement is either cyclic, and termed 'reversible moisture movement', or a once-off event at the start of the life of the component, which is termed 'irreversible movement'.

## Irreversible moisture movement

Irreversible moisture movement is the consequence of once-off change in baseline moisture content of a material or component as it dries out or absorbs moisture to harmonize with the ambient moisture level. According to the season when the building is constructed, the drying-out period is

*The Technology of Building Defects.* Dr John Hinks and Dr Geoff Cook.
Published in 1997 by E & FN Spon, 2–6 Boundary Row, London SE1 6HN, UK. ISBN 0 419 19770 2

variable and so too is the timing of the initial appearance of any directly associated faults. Irreversible movement is usually more extensive than reversible movement.

In the case of clay bricks there is a period of moisture absorption as they adjust from their post-firing dryness; half of this occurs in the first week following firing, and the remainder is usually complete in approximately three months. Expansion problems with bricks incorporated into a structure prematurely are therefore possible. In contrast, concrete exhibits irreversible shrinkage as it undergoes its initial drying out.

## Reversible moisture movement

Whilst irreversible moisture movement may involve shrinkage or expansion, reversible moisture movement involves both shrinkage and expansion in a recurrent cycle. Direct wetting or changes in the ambient moisture content produce swelling in the component material as it absorbs moisture in an attempt to reach equilibrium with the ambient surroundings. A corresponding shrinkage occurs when the ambient conditions change and become drier, and the material sheds moisture content to the surroundings.

Expansion stresses may be contained within the component or element or may break out as distortion or physical displacement, for example oversailing of brick parapets. The shrinkage cycle may produce tensile cracking in a component or element at a weak point. Note that oversailing displacement will not usually reduce with changes in moisture content; however shrinkage cracking may temporarily close up upon rewetting of the structure. These phenomena often coincide with reversible thermal movement and analysis of movement needs to be sensitive to the interactive dynamics which may occur. Symptoms of moisture and thermal movement may be superficially similar to those arising from substructure or ground movement.

## Symptoms of movement

The scale of movement and its recurrence are key factors in moisture movement, together with the relative intensity of the moisture-induced stresses compared with the intrinsic strength of the material.

The integrity and scope for deformation in the component or element, and/or the relationship between the affected component(s) and the remainder of the surrounding structure determine the symptoms of the defect. Visible defects arise where there is an intolerance of relative movement between components (usually expansion) within the design and/or construction detailing. Bulging or distortion may occur under moisture-induced expansion, tensile cracks under shrinkage. Where irreversible shrinkage occurs, or the reversible movement cycle proceeds into the shrinkage phase, cracking symptoms may appear where tensile stresses exceed the strength of the component and are released by splitting. It is therefore realistic to expect such cracking to alter in size with changes in ambient conditions (and therefore the phase in a reversible movement cycle).

Stresses which would otherwise be tolerable may create visible defects in cases where the rate of change in ambient conditions is not matched by the

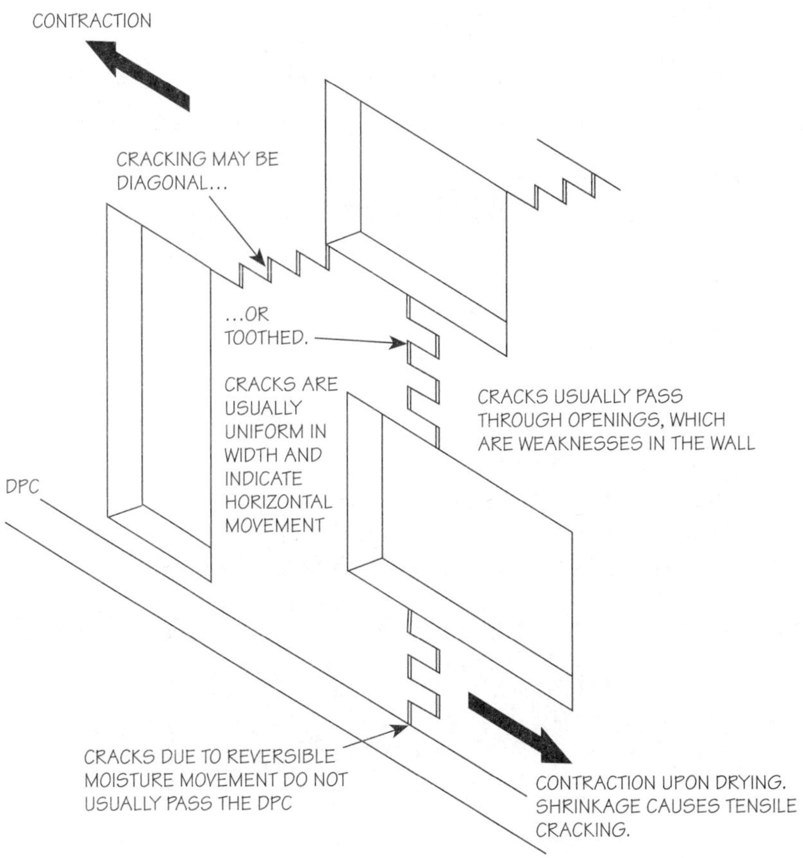

**Fig. 6.1.** Cracking due to moisture movement.

rate of moisture response in the material. In particular, the climatic variability in the UK may mean that components get out-of-phase with the climatic changes, as the drying-out period for materials may exceed the dry period between bouts of wet weather, leading to frequent cyclic moisture movement or chronic saturation associated with intrinsic stresses.

The exact form of symptom associated with moisture movement will depend upon the scale of the component and element, modified by its structural and other physical relationships with its surrounding components and elements. Short elements or components will have less scope for accommodating the displacements occurring to other local components, and so may exhibit defects caused by the movement of the larger, attached component. A similar effect on the location of symptoms tends to occur with components or elements that have dramatically differing stiffness or mass.

Very stiff or structurally-constrained components may retain stresses internally or partially release them by deforming, by bulging for example. Components or elements which are relatively free to move, perhaps by being the predominant component at an interface, may exhibit oversailing or a variety of forms of tensile cracking associated with shrinkage. Accordingly, one of the classic interface defects associated with in-situ concrete-framed construction and brick infill panelling arose where the irreversible

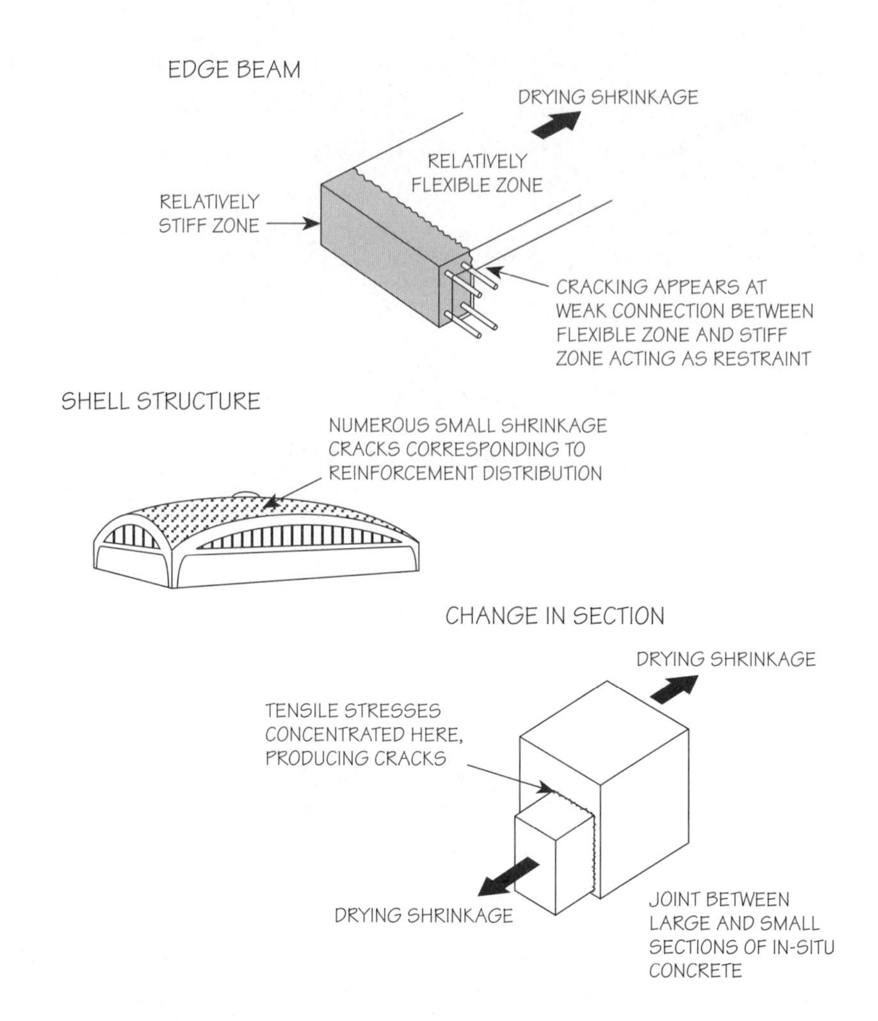


**Fig. 6.2.** Examples of drying shrinkage cracking to in-situ concrete frame.

expansion of the brickwork coincided with irreversible drying shrinkage of the surrounding concrete frame. Inadequate movement allowances led to bulging in the brick panelling and/or cracking at the corners of the frame. This was further exacerbated by reversible thermal and moisture movement in the brick panelling.

Cracks will follow the lines of least resistance, such as the joints between components, or sites of weakness within the component itself. In elements such as loadbearing masonry walls, which will tend to tolerate vertical movement well but horizontal movement poorly, shrinkage cracking may appear in the form of vertical cracks.

Since the areas of least structural resistance are usually around openings, then this is another likely area for movement cracking. Other areas where the resistance to movement is relatively low are around the tops of walls. Here the self-weight of the components is at a minimum and so too is the resistance to displacement.

Note that it is relatively unusual for moisture shrinkage cracking to cross the damp-proof course, since the structure below the DPC retains a more consistent moisture content level. This zone of the building also changes in moisture content more slowly, and is usually damper anyway. In addition, the modern DPC creates a good slip membrane within the wall, and the low resistance to movement allows the localized movement of the walling to release stresses.

Inside the building, the loading and self-weight of structural components and elements may be relatively light. Displacement will therefore be quite likely. The internal structure should not be subject to the same degree of cyclic changes in ambient moisture content as the external envelope; however, irreversible moisture movement may still be problematic.

## Revision notes

- A natural and fundamental cause of building defects.
- Can operate in isolation or in combination with other causes of building defects, including thermal movement.
- Extent of moisture movement and speed of response depend upon permeability of the material.
- Reversible moisture movement is cyclic and driven by the permeable material harmonizing its moisture content with the surroundings.
- Reversible movement may create faults which alter in scale cyclically. This is not always the case (e.g. oversaturated brickwork).
- Irreversible moisture movement is a one-off event at the start of the life of the building.
- May be irreversible shrinkage (e.g. concrete) or expansion (e.g. brickwork).
- Expansion and contraction stresses in materials may be contained or may appear as breakage or distortion.
- Cracking will follow the line of least resistance.
- It is unusual for moisture-related shrinkage to cross the DPC.

## Discussion topics

- Discuss the distinction between reversible and irreversible causes of moisture movement in buildings, and how the symptoms of such movements may be distinguished.
- Explain how moisture movement and thermal movement can combine to produce defects. How are the two movement causes distinguishable in defects?

## Further reading

Addleson, L. (1992) *Building Failures: A Guide to Diagnosis, Remedy, and Prevention*, 3rd edn, Butterworth-Heinemann.

Alexander, S.J. and Lawson, R.M. (1981) *Design for Movement in Buildings*, CIRIA Technical Note 107.
BRE (1975) *Principles of Modern Buildings*, 3rd edn, Vol. 1, Building Research Establishment.
BRE (1977) *Principles of Joint Design*, Digest 137; Building Research Establishment.
BRE (1979) *Estimation of Thermal and Moisture Movements and Stresses*, Digests 227, 228, 229, Building Research Establishment.
Bryan, A.J. (1989) *Movements in Buildings*, R3, CIOB Technical Information Service.
CIRIA (1986) *Movement and Cracking in Long Masonry Walls*, CIRIA Practice Note Special Publication 44.
Cook, C.K. and Hinks, A.J. (1992) *Appraising Building Defects: Perspectives on Stability and Hygrothermal Performance*, Longman Scientific & Technical, London.
Everett, A. (1975) *Materials*, Mitchell's Building Series, Batsford.
Morton, J. (1988) *Designing for Movement in Brickwork*, BDA Design Note 10.
Rainger, P. (1983) *Movement Control in the Fabric of Buildings*, Mitchell's series, Batsford Academic and Institutional.
Ransom, W.H. (1981) *Building Failures: Diagnosis and Avoidance*, E.&F.N. Spon, London.

# 6.2 Thermal movement

## Learning objectives

- The mechanism of thermal movement in building materials and components should be understood.
- You should also appreciate the role of stiffness and restraint in the manifestation of thermal movement defects.

All building materials experience thermal movement; however, the expansion coefficient varies between materials and therefore the actual movement and its significance for the building also vary.

Thermal movement is a reversible phenomenon. A background thermal movement occurs with seasonal changes; however, rapid, extreme thermal gains and losses are possible, especially with the UK climate, and therefore diurnal change can be more significant for the building. Rapid changes in temperature can create problematic stresses in the building, focused at connection points and/or at joints with insufficient movement tolerance.

A number of factors affect the amount of thermal movement occurring in a component or element. An important driver is the exposure of a component to temperature instability or temperature differential such as occurs with solar gain (and shading).

The mechanism of failure due to thermal movement in materials depends on the rate of change and the differential movement between components. Hence dark-coloured surfaces (which exhibit greater heat gain than light-coloured surfaces, and correspondingly greater thermal movement) are particularly prone to thermal movement. Dark surfaces can reach 140°C, compared with 80°C for otherwise similar light surfaces. Since it is the upper end of the temperature range which produces the significant differences in movement, components with dark surfaces are particularly prone to thermal movement.

The dimensional size of the component will influence the overall physical expansion or contraction which requires to be accommodated, hence larger components/elements produce greater tolerance problems.

Another important factor is the thermal capacity of the component. Highly insulated components will hold their surface heat and greater extremes of temperature will be experienced. As a result of the presence of the thermal insulation, temperature differentials are created across the thickness of the component, producing bowing. Composite structures such

*The Technology of Building Defects.* Dr John Hinks and Dr Geoff Cook.
Published in 1997 by E & FN Spon, 2–6 Boundary Row, London SE1 6HN, UK. ISBN 0 419 19770 2

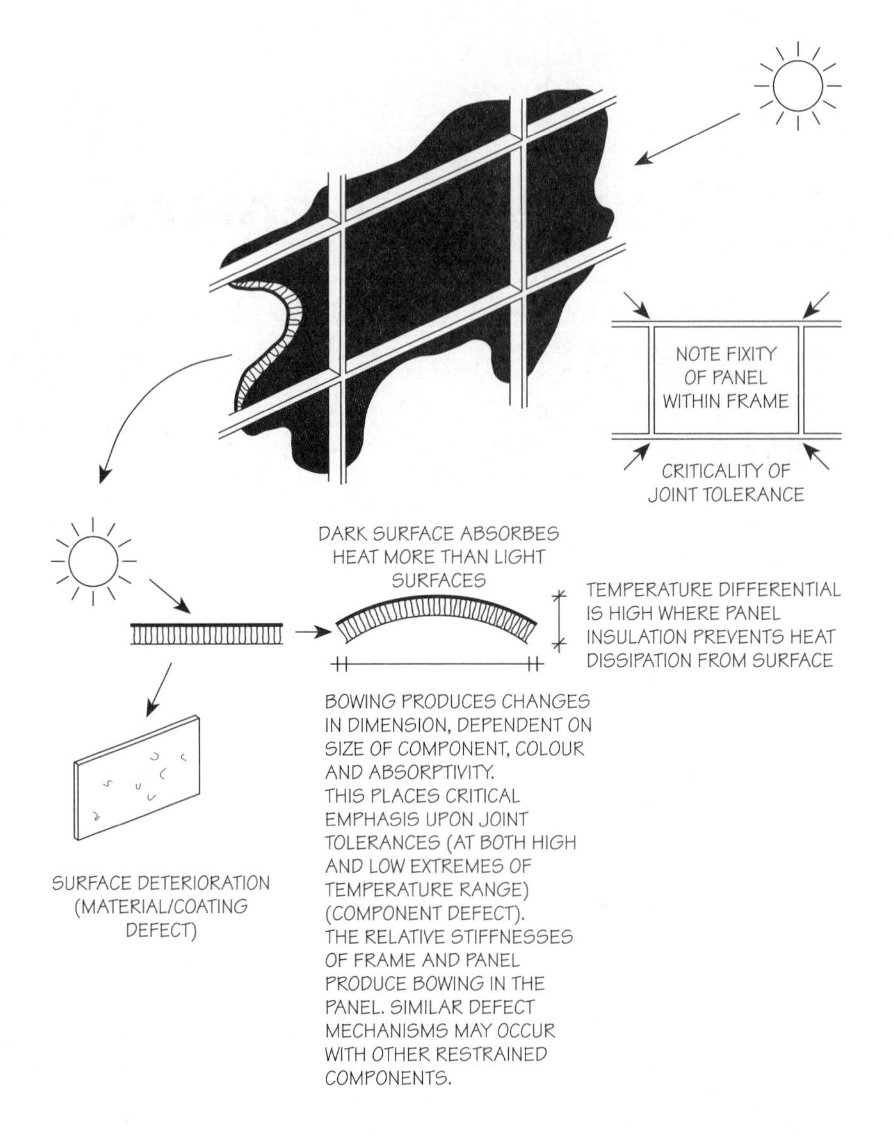


**Fig. 6.3.** Thermal movement.

as curtain walling may combine stiff frames of low thermal expansion coefficient with dark panels of relatively high thermal expansion. Careful design of the joint tolerances is needed to accommodate the potentially significant relative movement during extremes of temperature.

The region where rigid and flexible components/elements meet is a particularly vulnerable area for thermal movement problems. Whilst materials with different modulus of elasticity may exhibit the same degree of physical movement, the stresses built up in them where they are under restraint will differ according to their modulus of elasticity. Hence stresses in restrained brickwork made from strong bricks will be greater than those in weak brickwork. As a result, the potential for internal damage to the component, the element or the structure is greater where components of differing stiffness meet. The weaker component will tend to exhibit symptoms of thermal movement which may originate as stresses in the stiffer component.

## Symptoms of movement

The symptoms of thermal movement include buckling, cracking and detachment. This usually occurs at points of restraint such as connections to frames or at joints between components/elements. In unrestrained or poorly restrained regions, for example parapet walls, the movement may be significant. Since expansion movement does not require the same degree of connectivity as contraction, the components tend not to return to their original position, so that cyclic thermal movement may create permanent displacement such as oversailing.

Thermal movement can compound or compensate for moisture movement. Where the two forms of movement occur in concert the disruption can be significant.

**Fig. 6.4.** Spalling of exposed edges of external mosaic adjacent to joints.

## Revision notes

- A phenomenon affecting all building materials, which respond to different degrees.
- A reversible effect, exhibiting seasonal and diurnal patterns, the latter may be more significant.
- Factors affecting the extent and impact of thermal movement include temperature range, differential temperatures, colour and composition of background, thermal inertia generally, dimensional scale and position of appropriate joints/tolerances, strength and rigidity of component and surrounding structures.
- Symptoms are cracking, blistering and displaced movement in local (weaker) components (including joints).
- Cyclic thermal movement may produce permanent movement.
- Can compound or compensate for moisture movement, according to circumstances.

## Discussion topics

- Identify the direct and exacerbating factors determining the amount of thermal movement in buildings. Produce a prioritized list of these factors creating thermal movement and estimate the scale of movement that can be expected.
- In what circumstances is moisture movement likely to combine with thermal movement? What are the usual symptoms of such combined movements?

## Further reading

Addleson, L. (1992) *Building Failures: a Guide to Diagnosis, Remedy, and Prevention*, 3rd edn, Butterworth-Heinemann.

Alexander, S.J. and Lawson, R.M. (1981) *Design for Movement in Buildings*, CIRIA Technical Note 107.

BRE (1977) *Principles of Joint Design*, Digest 137, Building Research Establishment.

BRE (1985) *External Masonry Walls: Vertical Joints for Thermal and Moisture Movements*, Direct Action Sheet DAS 18 (February 1985), Building Research Establishment.

Bryan, A.J. (1989) *Movements in Buildings*, R3, CIOB Technical Information Service.

CIRIA (1986) *Movement and Cracking in the Long Masonry Walls*, CIRIA Practice Note Special Publication 44.

Cook, G.K. and Hinks, A.J. (1992) *Appraising Building Defects: Perspectives on Stability and Hygrothermal Performance*, Longman Scientific & Technical, London, pp. 260, 265–282.

Everett, A. (1975) *Materials*, Mitchell's Building Series, Batsford.

Morton, J. (1988) *Designing for Movement in Brickwork*, BDA Design Note 10.

Rainger, P. (1983) *Movement Control in the Fabric of Buildings*, Mitchell's Series, Batsford Academic and Institutional.

# 6.3 Joints and sealants

## Learning objectives

- You should be able to recognize the range of defects which can occur with movement joints, mastics and sealants.
- You will be able to identify the causes of these defects, which are associated with inadequate dimensional and material factors.
- You will appreciate the specific failure mechanisms of joints between concrete panels.

## Joints generally

Since all building materials have the potential for movement, a lack of movement accommodation can cause a range of defects. The movement can be due to changes in loading, thermal or moisture conditions of components or materials. Although it is possible to calculate the total range of movement to be accommodated, where this fails to include all of the factors which influence movement, then the joint may fail. Expansion joints may have insufficient width, contraction joints excessive width and general movement joints an inability to accommodate either.

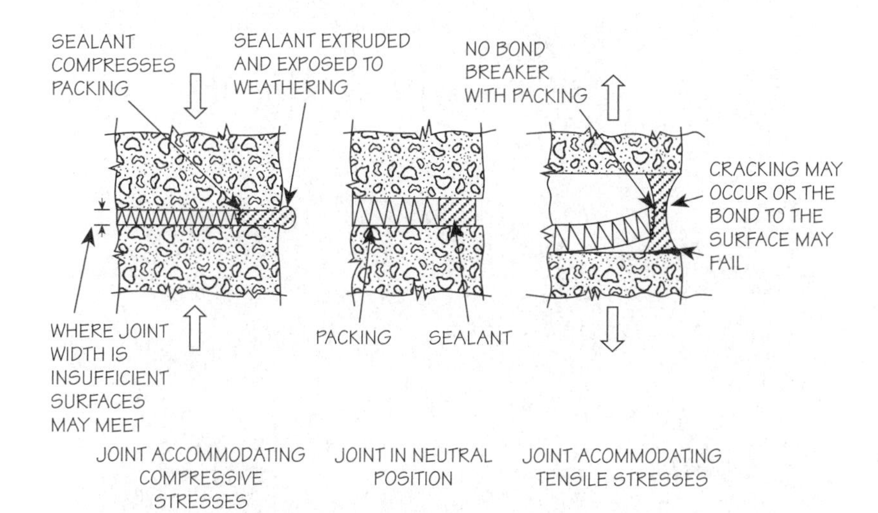


**Fig. 6.5.** Joints failing to accommodate movement.

*The Technology of Building Defects.* Dr John Hinks and Dr Geoff Cook.
Published in 1997 by E & FN Spon, 2–6 Boundary Row, London SE1 6HN, UK. ISBN 0 419 19770 2

The joint width must be adequate to accommodate the required expansion and contraction of the element or component on either side of the joint. In order to account for the practical aspects of joint formation the tolerances of the structural element and the construction process must also be included. In general the tolerances for factory-made components are likely to be lower than those formed in situ. This enables a maximum and minimum joint width to be determined. The 'movement accommodation factor' expresses the movement as a percentage of minimum joint width; typically this value is 20%. With closed systems of construction the tolerances should apply to all elements and assist in joint design. The joints between in-situ and factory-made elements, since each may have different tolerances, are more problematic.

**Fig. 6.6.** The precast concrete panel joints are poorly aligned and of differing widths.

Even where the movement range has been correctly assessed the joint may still fail because the sealant is subjected to either excessive tensile or excessive compressive stress. It is suggested that to avoid transfer of excessive compressive stress to the structure, compression joints 12–15 mm wide can be considered minima for brick or precast concrete cladding panels fixed to a reinforced concrete structural frame. Since the life of the sealant material is unlikely to exceed that of the materials, replacement is to be expected. A failure to appreciate the need for replacement can result in a defectively designed joint.

The inclusion of movement joints can be a potential source of problems as well as offering a solution. This is particularly evident with the industrialized building systems of the late 1950s and 1960s. Where the joints fail to accommodate movement the integrity of the surface will break down. In the case of external joints they may not remain weatherproof and for internal joints appearance or environmental performance may be affected.

Traditional lime mortar joints between masonry units had the ability to accommodate the irreversible expansive movement of new clay bricks as well as their long-term reversible movement. The development of cement

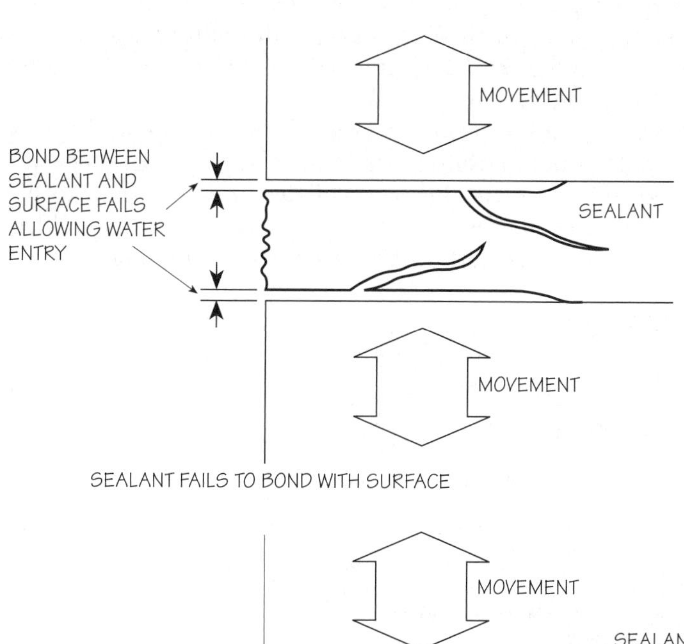


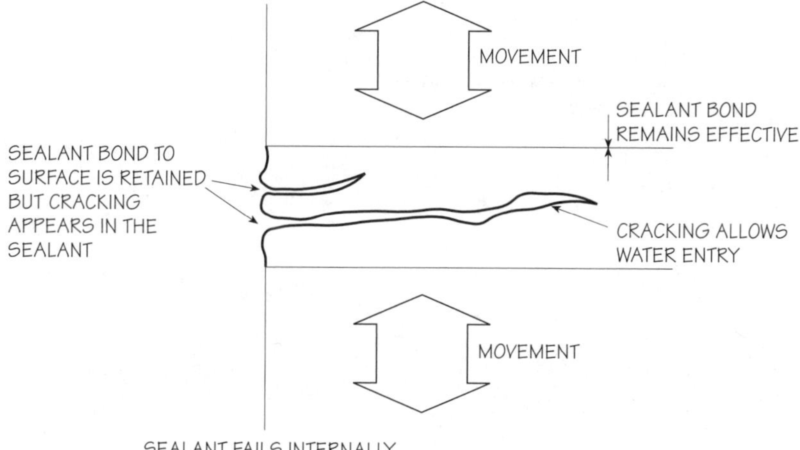


**Fig. 6.7.** Failure mechanisms of sealant. (Reproduced from G. Cook and A.J. Hinks. *Appraising Building Defects*, Addison Wesley Longman, 1992.)

mortars has tended to concentrate movement effects and demands the use of expansion joints.

## Movement joints – weather resistance

There are several types of movement joint and each type can fail to provide weather resistance, in a manner dependent upon the characteristics of the joint. The butt joint may be sealed with a mastic sealant. This places the mastic towards the front of the joint where weathering is most severe. UV degradation of sealants can cause embrittlement and the adhesion between mastic and surface can break down where the joint is exposed. Where joints have been replaced they may fail where the new sealant cannot effectively bond with the abutting materials. The remains of the original sealant may form a thin blocking film.

Small cracks can occur between bricks and the mortar joint which may allow water penetration. Gaskets, preformed joint fillers, can also be used for butt joints. Since they rely on the compressive stress from the surrounding surfaces to retain their integrity, where this is variable, or becomes tensile, then joint seals may open. The gasket material touches, rather than adheres to, the surrounding surfaces and where surface textures are variable then contact may be similarly variable.

The drained vertical joint used in precast concrete cladding panels incorporated a baffle which, although difficult to replace and occasionally missing altogether, relied on the integrity of the intersection with a horizontal flashing to provide weather resistance. Where the horizontal flashing was not dressed over the upstand of the lower panel, bringing the baffle to

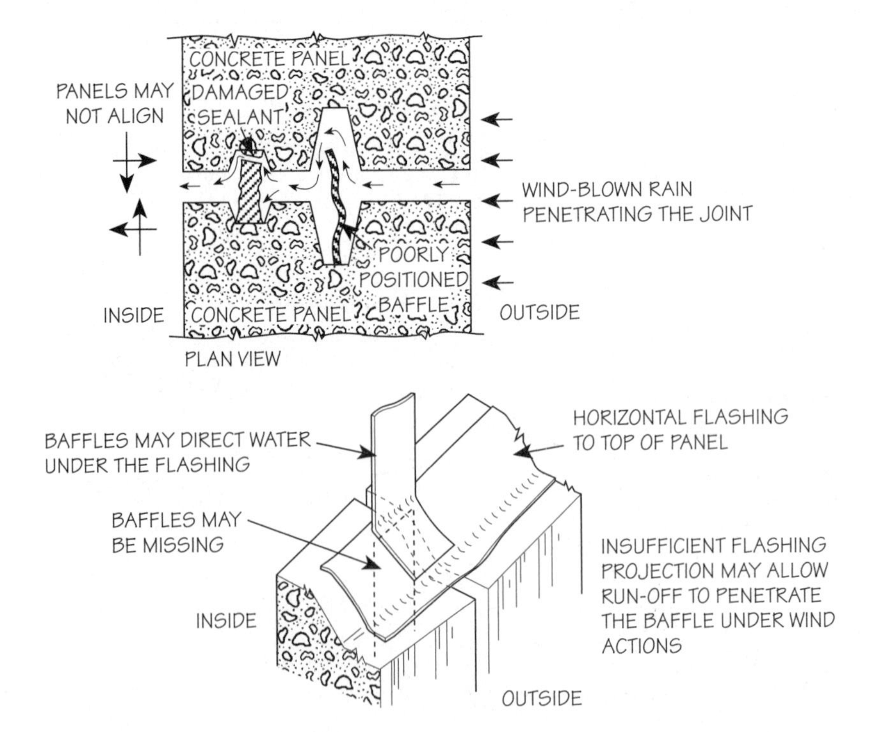


**Fig. 6.8.** The baffle–flashing joint.

the front of the panel, or did not provide adequate cover to the area behind the baffle, then it may fail. The sealant or gasket joint behind the baffle can fail, producing a joint which is no longer airtight and weather resistant. Sealants which are likely to be in contact with three surfaces can fail where a bond breaker has not been provided.

In tall buildings the wind flow pattern is complex and because wind speeds tend to increase with height the weathering effects are extreme. Rain can move vertically upwards and horizontally; in this way, the integrity of all sections of the jointing system is examined. The failure zone, which allows water to enter the building, can be difficult to locate owing to the configuration of panels and joint sealants and flashings.

### Revision notes

- The lack of movement accommodation in building elements and components can cause a range of defects. Movement may be due to loading, thermal or moisture conditions.
- The joint width must accommodate the maximum expansion and contraction expected. This should include dimensional tolerances associated with either factory or site conditions.
- Sealant replacement, which is very likely to be required, may be impossible or difficult. Similarly the drained vertical joint may have missing baffles or the associated flashings may fail to provide weather resistance.

---

## Discussion topics

- Compare the movement characteristics of three construction materials and their constructional tolerances.
- Describe the failure mechanisms of movement joint sealants.
- Explain why a typical 'movement accommodation factor' is 20%.
- Illustrate the possible water paths through precast concrete cladding panels where drain joints have been used.
- 'Traditional construction was more able to accommodate movement than the newer types of construction which have made specific allowance to accommodate movement.' Discuss.

---

## Further reading

BRE (1974) *Estimation of Thermal and Moisture Movements and Stresses: Part 1*, Digest 227, Building Research Establishment, HMSO.

BRE (1979) *Estimation of Thermal and Moisture Movements and Stresses: Part 2*, Digest 228, Building Research Establishment, HMSO.

BRE (1979) *Estimation of Thermal and Moisture Movements and Stresses: Part 3*, Digest 229, Building Research Establishment, HMSO.

BRE (1983) *External Masonry Walls: Vertical Joints for Thermal and Moisture Movement*, Defect Action Sheet 18, Building Research Establishment, HMSO.

BRE (1985) *External Walls: Joints with Windows and Doors – Application of Sealants (Site)*, Defect Action Sheet 69, Building Research Establishment, HMSO.

BRE (1990) *Joint Sealants and Primers: Further Studies of Performance with Porous Surfaces*, Information Paper 4/90, Building Research Establishment, HMSO.

Cook, G.K. and Hinks, A.J. (1992) *Appraising Building Defects: Perspectives on Stability and Hygrothermal Performance*, Longman Scientific & Technical, London.

Everett, A. (1975) *Materials – Mitchells Building Construction*, Longman Scientific & Technical, London.

PSA (1989) *Defects in Buildings*, HMSO.

Ransom, W.H. (1981) *Building Failures: Diagnosis and Avoidance*, 2nd edn, E. & F.N. Spon, London.

Richardson, B.A. (1991) *Defects and Deterioration in Buildings*, E. & F.N. Spon, London.

Taylor, G.D. (1991) *Construction Materials*, Longman Scientific & Technical, London.

# 6.4 Thermal inertia

## Learning objectives

- You should understand the effect thermal inertia has on the thermal response of building materials.
- Macro- and Micro-scale effects of differences in thermal inertia produce a range of responses in buildings.

Thermal inertia is an important controlling characteristic of thermal movement. It is particularly relevant in the context of differential thermal movement of partially shaded buildings. In the context of the mass of a building, it also affects the time required for the building to adapt to ambient temperature changes.

On a micro-scale, this means that the relative movement of building components and elements is dependent, in part, on their thermal mass and exposure to temperature change. Thermally lightweight components comprising materials with a low thermal capacity will change temperature (and hence dimension) relatively rapidly. Components that have a high level of thermal inertia will change temperature and dimension less dramatically.

Taken on a macro-scale, high-mass elements of buildings, which proportionately take a larger amount of energy to heat up than smaller-mass components of a similar material, will tend to take longer to adapt to changes in ambient temperature. The climatic variability in the UK is therefore a significant driver, and massive buildings with high thermal inertia can become out-of-phase with rapidly cycling ambient temperatures. This produces a time lag phenomenon in massive buildings, as they lag behind any relatively rapid climatic cycling. This phenomenon mostly appears in the older, heavyweight stock, but also includes the heavyweight concrete systems construction. Defects may occur in such buildings where the internal climate becomes out of phase with changes in the external climate and/or the building fabric, or alternatively with the cycles of usage. With intermittent use of such buildings, the consequence of relative changes of the internal humidity is frequently surface condensation. As occupiers return to a cold building and rapidly warm up the internal air with central heating, for instance, the structure lags behind and provides the cold region for the surface condensation of water vapour. Coupled with the occurrence of poor ventilation, the problem may become chronic.

*The Technology of Building Defects.* Dr John Hinks and Dr Geoff Cook.
Published in 1997 by E & FN Spon, 2–6 Boundary Row, London SE1 6HN, UK. ISBN 0 419 19770 2

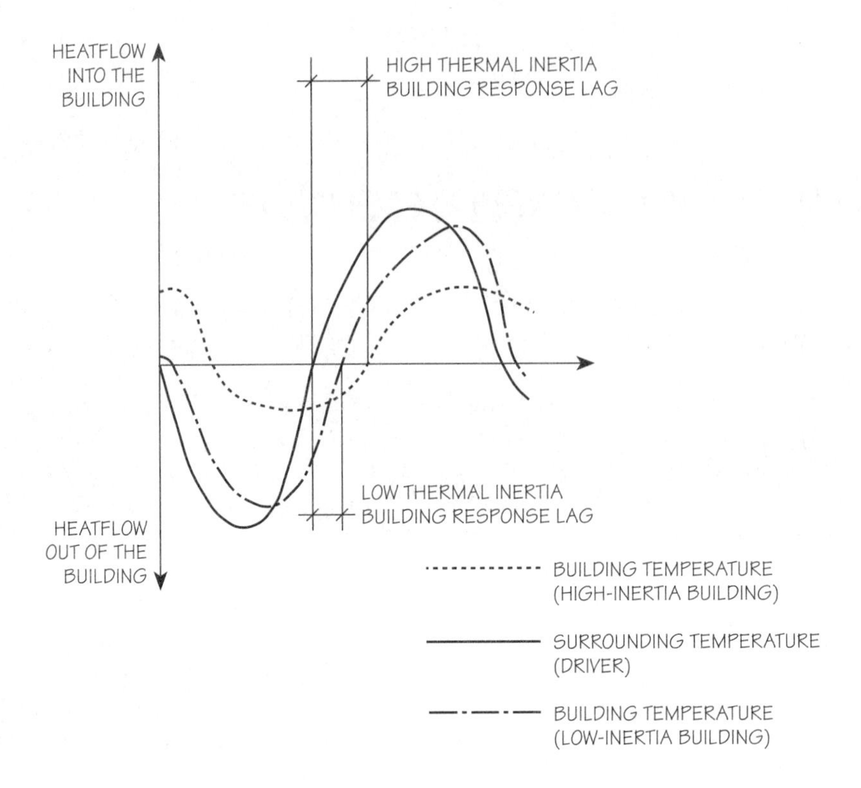


**Fig. 6.9.** Time lag and thermal inertia. Buildings with low thermal inertia exhibit less response lag than those with high thermal inertia.

Clearly there is also a profound relationship with thermal insulation levels. Modern lightly insulated buildings may be of low thermal inertia, and exhibit little or insignificant lag in responding to thermal change. They therefore respond better to intermittent usage. In such cases the building elements and components will be much more mobile. The building also provides a minimal buffering effect against changes in climate.

## Revision notes

- A controlling factor in thermal movement.
- Particular relevance in the UK because of variability in the climate.
- Buildings with high thermal inertia react relatively slowly to change in temperature. Temperature and dimensional changes in the component/element will be relatively moderate and will lag behind ambient changes.
- Buildings which are out of phase with the initial ambient conditions may suffer condensation.

## ■ Discussion topics

- Discuss what part thermal inertia plays in the occurrence of defects associated with (a) thermal movement and (b) condensation.
- On the basis of this, identify the most and least vulnerable constructional forms.

## Further reading

Addleson, L. (1992) *Buildings Failures: A Guide to Diagnosis, Remedy, and Prevention*, 3rd edn, Butterworth-Heinemann.

Alexander, S.J. and Lawson, R.M. (1981) *Design for Movement in Buildings*, CIRIA Technical Note 107.

BRE (1975) *Principles of Modern Buildings*, 3rd edn, Vol. 1, Building Research Establishment.

BRE (1977) *Principles of Joint Design*, Digest 137, Building Research Establishment.

BRE (1979) *Estimation of Thermal and Moisture Movements and Stresses*, Building Research Establishment.

Bryan, A.J. (1989) *Movements in Buildings*, R3, CIOB Technical Information Service.

CIRIA (1986) *Movement and Cracking in Long Masonry Walls*, CIRIA Practice Note Special Publication 44.

Cook, G.K. and Hinks, A.J. (1992) *Appraising Building Defects: Perspectives on Stability and Hygrothermal Performance*, Longman Scientific & Technical, London, pp. 260, 265–282.

Everett, A. (1975) *Materials*, Mitchell's Building Series, Batsford.

Morton, J. (1988) *Designing for Movement in Brickwork*, BDA Design Note 10.

Rainger, P. (1983) *Movement Control in the Fabric of Buildings*, Mitchell's series, Batsford Academic and Institutional.

Ransom, W.H. (1981) *Building Failures: Diagnosis and Avoidance*, E. & F.N. Spon, London.

# 6.5 Paintwork

## Learning objectives

- You should understand the range of causes of defects in paintwork, and be able to distinguish between faults in the paint coating itself, and those in the background.
- Note that background faults may be causal or consequential of paint defects.

Defects in paintwork arise from many causes, mostly in combination, but the main culprits are exposure to sunlight (especially so with dark coatings) and rain. Pollution and a range of instabilities of the background can also cause problems. Deficient paintwork may also indicate deeper problems. Paintwork relies heavily on thorough and appropriate maintenance.

Deferred maintenance allows the protection to diminish, thus leading to greater damage to the background. The cost of renovatory work also increases, since paintwork in an advanced state of decay is more difficult to prepare and remedy.

Poor paint adhesion occurs if the background is damp during application. Painting in high-humidity conditions will affect the quality of the coat and its eventual durability.

Water-related problems will arise during the life of the coating if condensation occurs behind the paint, or if surface defects allow water to penetrate the surface coating on a large scale.

## Timber backgrounds and paint defects

The principal function of external paintwork is to control the entry of water into timber. Wood alters its moisture content to approach equilibrium with the environment and this produces disruptive changes in size. The changes are not the same in all directions: more expansion occurs across the grain than along it.

If timbers are continually changing in moisture content this can give rise to deterioration caused by splitting and distortion. The grain may become obvious and unsightly, too. The extent of moisture-related movement that timber undergoes depends on the particular wood. However, this can be significant enough to cause problems with poor-adhesion paints.

*The Technology of Building Defects.* Dr John Hinks and Dr Geoff Cook.
Published in 1997 by E & FN Spon, 2–6 Boundary Row, London SE1 6HN, UK. ISBN 0 419 19770 2

Because of these factors, extensibility without cracking (so the paint skin remains intact) is an important property. Otherwise, the wood and the paint move differentially, causing flaking and an incomplete coating. Once the coating is deficient moisture control is lost, and the situation worsens rapidly, until repainting becomes essential to preserve the wood.

The way in which the paint allows water vapour to pass through it will determine how naturally the timber can move, and how quickly it can approach equilibrium with the environment. An impermeable coating allows complete control of the moisture content of timber, which could theoretically eliminate problems with rot if applied to completely enclose dry timber. However, the completeness of the skin cannot be assured. There is a distinct likelihood that any moisture entering the timber will become trapped beneath the coating. While rot and possibly frost damage may occur in the wood, the pressure of entrapped water will cause the remaining paint to blister or peel off, so worsening the situation.

In contrast, a paint that is too permeable will control the moisture content of the timber poorly (the problem with traditional stains) and allow excessive movement.

Clearly then, correct permeability is an important function of paint. The permeability of a coating varies widely from the low-solids stains (see later) to the traditional finish of several undercoats and a gloss finish over a primed surface.

There are a number of specific categories of paint defect, described below.

## Flaking

'Flaking' occurs with the loss of adhesion between the paint layer and substrate, so that small particles break away. Paints bond less effectively to non-porous surfaces, and some oily hardwoods such as teak are troublesome. Flaking and other bond-related faults occur more easily with such surfaces. All or some of the paint layers may be affected. Flaking is usually an external problem, which tends to arise with hard-drying paint types, or any paint applied to a friable surface. Timber backgrounds can also promote early flaking of paint because of trapped moisture, and this problem is particularly likely at the joints which will experience the greatest movement.

Timber that is highly grained or well weathered is likely to produce an unstable background. As differential movement occurs across the grain or within the weathered timber this will set up surface stresses which destroy the integrity of the painted surface. Layers may flake off following the pattern of the grain, so illustrating the nature of the stresses. Clearly this vulnerability in the background will lead to rapid deterioration if unattended.

## Bubbling

The formation of small blisters or bubbles in the wet paint surface can produce pinholing of the surface. Energetic mixing of the paint or vigorous

**Fig. 6.10.** Flaking and cracking of paint film following the grain of the timber background. (University of Reading.)

application froths the paint, and small bubbles are created in the wet surface. These dry more rapidly than the surrounding film and the bubbles become trapped in the finished coating.

Bubbles in the painted surface can also appear on hardwoods. These are the result of trapped moisture, air or other gases, or from resins and crystalline salts. This 'blistering' affects the dried film, and is worse with dark-coloured coatings which reach higher temperatures.

The problems with flaking and detachment from metallic surfaces may be related to residual mill scale, a galvanic coating or simply excessive thermal movement in the background.

## Peeling

'Peeling' is a large-scale form of flaking produced by differential movement. It can be caused by the paint layers on their own, as a reaction to other (background) paints or between the paint coating and the substrate. This is distinct from 'cissing', where layers 'roll off' the background. Cissing is produced by incomplete wetting of the background, leading to patchy adhesion.

## Checks and crazing

Poor-quality paints may develop 'checking' in the form of 'V' or 'crows-feet'-shaped cracking. These cracks are very fine and whilst these do not immediately threaten the protective properties of the film, if left unattended will lead to 'crazing' (or 'alligatoring'). Crazing defects arise most frequently with paints based on or containing bitumen. They occur when hard coatings are applied to soft backgrounds. As these dry they shrink and distort the background. Adding excessive driers can produce the same effect. Patterned cracks form in the dried paint but the defect does not always penetrate the film completely. The cracks are raised above the surface producing an alligator-skin effect.

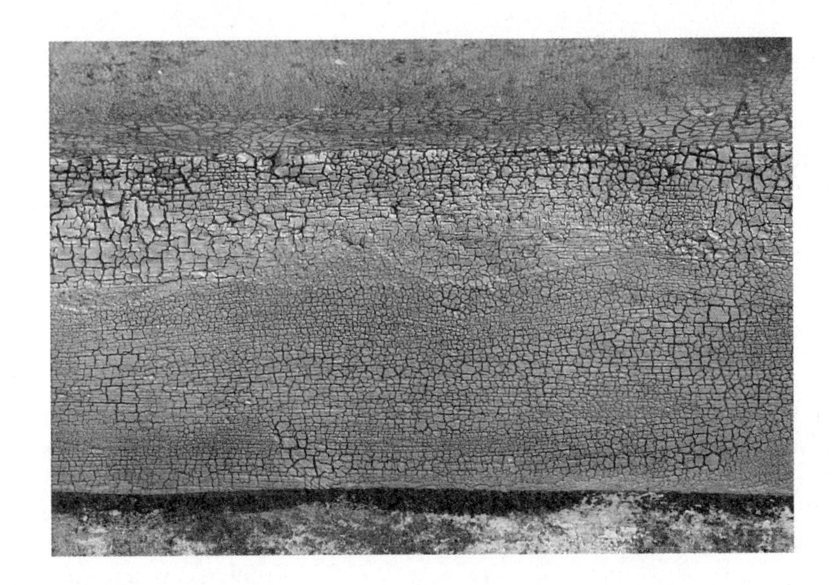

**Fig. 6.11.** 'Alligatoring' of a paint film. (University of Reading.)

## Chalking

Mildly eroding surfaces exhibit 'chalking'. Simple chalking may not significantly affect the protective properties of the film, and is characteristic of older or over-thinned paints. The mechanism commences when the binding agent is oxidized, a process which may be accelerated by sunlight. Interior paints used in exposed locations are susceptible, as are highly pigmented spirit-borne paints. The unbound pigmented filler lies loose on the surface, spoiling the appearance. The effect is an easier paint to maintain and restore than those which have cracked or chipped.

Highly porous background surfaces which were incorrectly primed produce similar effects as the background sucks moisture out of the drying paint film. Paints which contain coarse pigment produce rough, flat surfaces. If these are abraded they 'polish up' to form a patchy appearance.

## Bittiness and pimpling

'Bittiness' occurs when dirt is incorporated into the surface from the atmosphere or dirty tools, or if the paint is not properly strained. Gloss surfaces show the problem most, flat-finish appearances are relatively tolerant of dirt. Other surface appearance defects arising during application include 'brushmarking' (or 'ropiness'). Severe brushmarking produces weaknesses in the paint film.

Where paints with poor flow characteristics such as high viscosity are sprayed, there is a tendency to produce surface pimpling. The problem is less frequent with brushing. This 'orange peel' effect may not be problematic unless the surface finish is critical.

## Blooming

This is a chemical response to the environment. It has a whitish appearance most noticeable on dark-coloured films, and occurs in damp atmospheres. It is characterized by a loss of gloss. Its occurrence is exacerbated by the presence of pollution. Oil-based paints are generally affected – particularly those with a high proportion of resin.

## Wrinkling

'Wrinkling' or 'rivelling' occurs as fine corrugations in the film shortly after paint application. It does not necessarily produce a fault in the covering quality, unless interfered with in the semi-dry state, but is unsightly and collects dirt. Uneven paint application will produce this defect. Edges of surfaces tend to collect thick paint coatings as the fluid paint gathers; these may run and rivel. Paint quality and skill in application are important instruments in preventing this. In extreme cases 'sagging' will occur. Paints which flow particularly easily can produce sagging, and their coating thickness must be carefully controlled to prevent deficiency.

## Colour variability

Colour variability is aesthetically problematic, but does not necessarily indicate any deficiency in protection. Incomplete mixing, or old or poor-quality paint can produce a variation in colour density or tone. 'Floating' may result, as the constituents separate out. Fading may even occur in the tin. There are a number of other potential causes aside from variability in the paint. Where the previous coating is visible through the overpaint, this is termed 'grinning through'. Incompatibility with previous coatings can cause a partial effect of bleeding. For example, bituminous paints are the most problematic bases, and can produce brown 'staining' through over-painting. Untreated knots will produce similar effects. Painting applied to ferrous metals may develop rust spots and stains once in place. Soluble salts in porous backgrounds will also affect paint films as they emerge from the substrate.

## Balling and saponification

Paints that do not dry out properly will remain 'tacky'. If the paint surface is rubbed down prematurely the paint may lift up and produce a 'balling' effect at the surface. This is difficult to cut down for recoating. Such problems arise also from lifting of the paint, produced by incompatibility with the substrate or undercoating layers.

The commonly cited problem of 'saponification' is the soapy deterioration of paint attacked by damp alkaline backgrounds. Walls containing cement, or timber stripped using an alkali paint remover may produce a softening or even liquefying of alkyd paints, or bleached patches in some emulsions.

## Specific faults with masonry paints

Faults with the product choice or background preparation are common sources of failures of masonry paints. Application to unsound backgrounds also causes problems.

Masonry paints have specific performance criteria to meet, including protecting the substrate from environmental pollution. It is likely that the substrate will have been painted in order to conceal a disintegrating surface, and unless the substrate is sound, the durability of the paint coating will be limited. Where cracks in brickwork have been concealed these may emerge firstly at mortar joints. A common cause of failure of these paints is moisture within the structure during painting or rising damp and condensation. All affect the quality of adhesion. As with other paints, separation of the substrate and paint can cause problems with water retention in the gap between the two, and this will accelerate any degradation. However, with masonry paints the nature of the substrate frequently makes their removal complicated.

The paints must be immune to the alkali in mortar and concrete and will have to resist any efflorescence, when salts in the burnt clays are brought to the surface. Only in prolonged and severe cases is it serious or more than an appearance problem. Porous paint films may allow the passage of the salt solutions without damage. It is possible that water which penetrates the paint through cracking will become trapped behind the coating, leading to prolonged dampness in the substrate.

## Other general comments

Some parts of the external structure are more vulnerable than others, and provide useful early symptoms. Rapid deterioration occurs where water can collect on surfaces (such as window sills), or becomes entrapped within joints or by bad design detailing. End-grain timber is especially vulnerable to water uptake: the rate can be several hundred times greater than that of lateral wood surfaces.

Sloping or horizontal surfaces collect water and suffer greater erosion and pollution damage than vertical ones. Several pollution and marine atmospheres all deteriorate painted surfaces more rapidly. Elevations facing south or west are affected by sunlight and wind most, both of which accelerate the ageing process.

Overpainting existing coatings rather than removing them can produce a range of problems – incompatibility, poor base surface quality or integrity, cleanliness or extent of preparation. Different types of paint may cause differential stresses, and the additional weight of the new coating may

**Fig. 6.12.** Peeling paintwork on a rendered wall. The leakage from the downpipe may have contributed to the paintwork failure. (K. Bright.)

overstress an old, friable base. Particular problems arise between old and new coatings of oil and water-based formulae.

The potential of the modern environment to damage coatings and building structure alike has imposed a greater criticality on paintwork. Durability is no longer just a function of the product, the preparation of the substrate and its application. Although there is now sufficient understanding to design paints on a chemical basis, the problems with the variability in substrate conditions and the local environment remain. Once deterioration to paintwork starts it can progress very rapidly. Frequent inspection for damage or general deterioration is advisable.

## Revision notes

- Defects have many possible causes. May occur in combination.
- Defects occur within the paint and in the background.
- Common causes include:
  - damp background, high porous backgrounds, dirty background;
  - dimensionally or chemically unstable background;
  - incompatibility between background and paint;
  - permeability of coating;
  - loss of adhesion;
  - poor paint quality, uneven application.
- Common symptoms include (probable causes in brackets):
  - flaking (loss of adhesion);
  - bubbling/blistering (mixing/application fault or moisture/resin entrapment);
  - peeling (differential movement);
  - cissing (patchy adhesion);
  - checks/crazing (paint quality);
  - chalking (eroding of paint surface);
  - bittiness/pimpling (incorporation of dirt);
  - blooming (chemical response to environment);
  - wrinkling (uneven paint application);
  - colour variability;
  - balling/saponification (poor drying out/alkaline reaction).
- Dark-coloured coatings can produce greater problems.
- Areas susceptible to water collection will generally exhibit faster paintwork deterioration. South or west-facing elevations are more vulnerable because of high exposure to sunlight and wind.

## Discussion topics

- Describe the range of visual defects that paintwork can exhibit.
- Discuss the causes of such defects, and the possible remedies.

## Further reading

BRE (1977) *Painting Walls. Part 2: Failures and Remedies*, Digest 198, Building Research Establishment (February)

BRE (1989) *External Masonry Painting*, Defect Action Sheet Building Research Establishment (July).

BRE (1990) *Painting Exterior Wood*, Digest 354, Building Research Establishment (September).

BRE (1995) *Maintaining Exterior Wood Finishes*, Good Building Guide 22, Building Research Establishment (August).

Hinks, A.J. and Cook, G.K. (1989) Defects in paintwork. *Building Today*, 21 September.

Snelling, J.E. (1996) *Painting and Decorating Defects: Cause and Cure*, E. and F.N. Spon, London.

# 6.6 External cementitious rendering

## Learning objectives

You should understand that:

- rendering defects occur within the coating itself and in the background;
- background defects may be a cause of consequence of faults in rendering;
- faults in rendering can mirror underlying defects and exacerbate them;
- there are general forms of defect in rendering.

## Introduction

Cement rendering can be an effective and durable external finish to buildings, but can suffer from a range of defects associated with its design and workmanship. There is an increased incidence of reported rendering defects, which could be related to a range of reasons beyond the defects attributable to construction design and operations, including the general awareness of the problem, the extended liability for construction work and the growth in the rendered building stock. The common forms of render defect are cracking and/or discoloration of the surface.

Where the render coat has developed defects as a result of faults in the underlying wall, this may create a wall which is doubly defective. Alternatively, a fault in the protective render coating can lead to accelerated secondary failure in the wall. This complicates the diagnostic process, and in these instances it is particularly important to eliminate the root cause of any defects before attempting remedial works.

There may be defects in the material, for instance dissolved salts and minerals in the water, which can directly affect durability. When working under winter conditions, any retarding effect of the dissolved salts could leave the render vulnerable to early frost attack. It is unusual, however, for render mortar to be attacked unless there is a high salt content. Hence vulnerable locations are around or below the DPC.

*The Technology of Building Defects.* Dr John Hinks and Dr Geoff Cook.
Published in 1997 by E & FN Spon, 2–6 Boundary Row, London SE1 6HN, UK. ISBN 0 419 19770 2

Defects may also arise in the background or as a result of incompatibilities between the render and the background. The mechanical bond between the two may have failed, owing to chemical or physical incompatibility. Movement (both temperature- and moisture-induced) as well as the irreversible inherent shrinkage that takes place in cementitious materials may combine to stress the interface bond. Renders which have a very high cement content will tend to exhibit abnormal shrinkage problems. Faults in the line and level of the background (which may be seen through some of the thinner-coated rendering systems) can produce attachment problems.

Loss of adhesion may also be due to renderings being applied to saturated walls or an incompatibility between the render and the background. The cause of this may lie with the render specification or the background stability. Frost and cold weather can also cause weakening of the rendered

**Fig. 6.13.** The banded cracking and detachment of the render is most marked where the gable parapet is exposed at the eaves.

coat. Dashing may fall off the render coat owing to the undercoat having a high suction. The tensile forces of dashing attachment can be significant, and weak backgrounds may not be suitable.

Loss of strength, powdering of the surface and general failure of the render may be due to the application of ready-mixed mortar beyond the recommended retardation period.

## Appearance defects

Plain sheets of cement render present an appearance of utilitarian dullness, even when sound, but are likely to craze and have a patchy appearance. Other (acquired) defects associated with rendering appearance are commonly located within the texture of the external face. Where this is deep and presents a series of non-self-cleansing surfaces to the weather, then discoloration can be rapid. Sulphate attack can also cause this general failure, although it tends to be localized.

## Cracking

Cracking in rendering may be associated with the breaking away of sections of the render. Breaking away increases the risk of rain penetration into the structure. By allowing the rain admittance to the back of the intact render

**Fig. 6.14.** Cracking patterns in rendering.

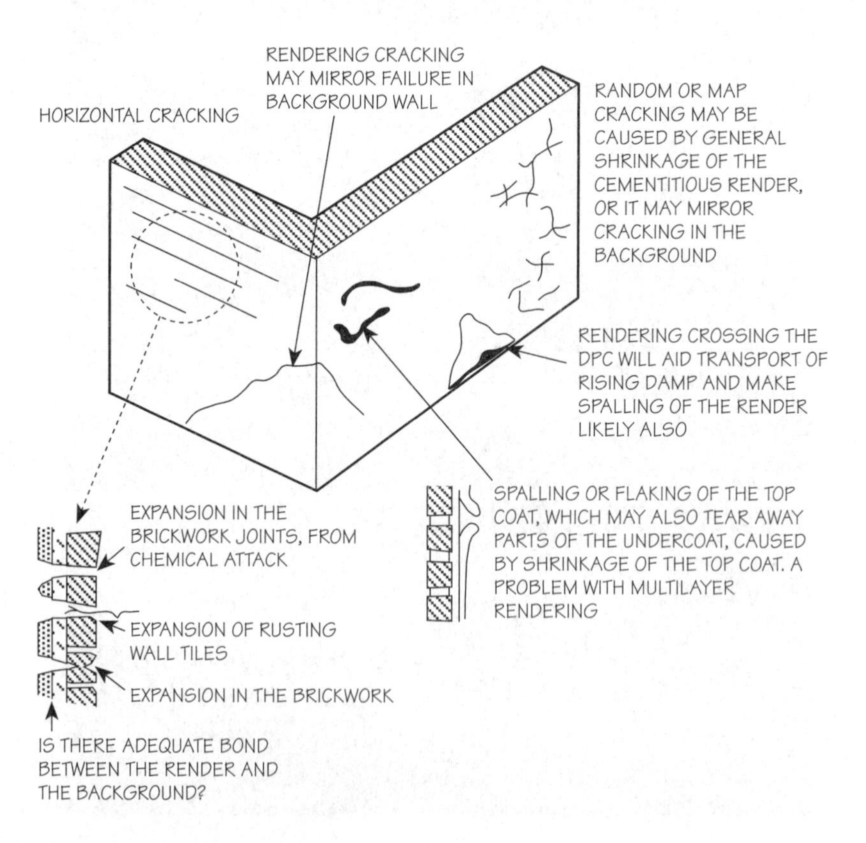

coat, any weakness in the bond between render and background or between different render coats will result in further defects and a vicious circle will occur, producing rapid failure.

Shrinkage cracks occur with the hydration of cementitious material, and are particularly marked where a rich mix has been used (1:3 or richer). Uneven concentrations of cement in mixed batches can create areas of render with high cement concentrations, and therefore differential shrinkage characteristics. Over-trowelling of the render can cause the cement matrix to rise to the surface, further increasing the risk of shrinkage. This form of failure is exhibited as map cracking.

Cracks associated with sulphate attack in the rendering may be general and without a particular pattern. Where they are predominantly horizontal this could indicate sulphate attack to the mortar joints in the background wall. As the joints crack with sulphate expansion, this is mirrored in the surface rendering. The advanced stages of sulphate attack or movement problems in the wall may eventually cause the detachment of large sections of rendering.

Do not immediately discount other causes of background cracking, including wall tie failure, which rendering may have been used to cover up. Regular patterns of cracks in rendering will usually indicate defects in the wall which are being carried through to the render. Cracks wider than 0.1 mm are liable to allow water penetration.

## Spalling

Small patches of spalling may develop where areas of the render have broken away from a backing brick wall. This may be due to a defect in the underlying brick. Older clay bricks can be of variable quality and underburnt bricks may have an increased soluble salts content, perhaps containing sulphates. In this case, they may also be generally weaker, since the vitrification process has not been fully achieved throughout the brick. The consequence is to reduce frost resistance, resulting eventually in detachment of the render coat.

## Large-scale spalling

May be due to loss of adhesion between the render and the background. Where water enters the interface between the remaining render and the wall, a progressive bond failure can occur. A free-standing sheet of render can be formed, which is held in place by the (minimal) inherent strength and stiffness of the render and its attachment to the background at random points. This can give rise to large-scale, and occasionally unexpected, failure.

## Flaking

In order to achieve the required thickness, or to mask surface imperfections in the wall, it is usual to apply render in several coats. Where subsequent

**Fig. 6.15.** The rendering may or may not be bridging the brick DPC. The run-off water from the wall and general weathering have started to detach the rendering from the wall.

coats are of higher cement content than the undercoats, there exists the possibility of differential shrinkage within the thickness of the multilayered rendering. The stresses generated by such shrinkage may be sufficient to exceed the relatively low tensile failure stress of the render/render interface. Any deficiency in the mechanical keying to the background will increase this possibility. This can lead to the final coat flaking away, sometimes taking with it some of the undercoat.

## Blistering

This is generally caused by an impervious paint film trapping the water vapour attempting to escape from the rendering. This may produce local areas of high moisture content, which may then be vulnerable to frost attack. Any faults in the rendering will enable moisture-induced cracking to commence, together with the other potential defects discussed above.

The typical wall defects which can cause cracking, such as tree roots and general subsidence, will directly affect rendering. The usual scope to disperse the small movements associated with these defects throughout the masonry joints cannot occur. Note also that rendering applied in areas of chronic damp and high exposure which fails will exacerbate the problem.

## Revision notes

- Defects can arise from the material itself; it can also exhibit defects in the underlying fabric.
- Common defect symptoms are cracking and surface discoloration.
- Defects in rendering may create or compound existing defects in the background, thereby complicating the diagnosis.
- Defects in the material arise from dissolved salts and minerals, or poor specification/workmanship, including the cement content.
- High cement content is one cause of shrinkage-cracking defects.
- Forms of failure include adhesion failure, major and map cracking in the render coating.
- Delamination may occur, with the outer coating of rendering flaking off from the basecoat.
- Failures in rendering which allow water penetration can lead to water entrapment and associated defects including increased susceptibility to frost attack in the render and background.

**Table 6.1** General defects and causes for cementitious rendering

| Defect | Cause |
|---|---|
| Cracking (random) | Where cracks are related to cracks in the wall<br>Shrinkage of rich mix<br>Excessive trowelling<br>Sulphate attack from background (see horizontal cracking) |
| Cracking (horizontal) | Sulphate attack<br>Where general, the source may be a clay brick background<br>Where localized, the source may be the mortar. |
| Areas detached in random pattern | Underburnt clay bricks |
| Flaking | Cement content of top coats greater than that of undercoats<br>Poor mechanical key<br>Shrinkage of the background |
| Blistering of paint film<br>(Where the back of the film is soft) | Water in the rendering<br>Saponification |

## ■ Discussion topics

- Describe the range of cracking defects that may appear in rendering.
- Explain and distinguish between the inherent defects which can appear in rendering and the symptoms of defects occurring within the substrate.

## Further reading

BRE (1983) *External Walls: Rendering*, Defect Action Sheet DAS 37 (October), Building Research Establishment.

BRE (1995), *Cementitious Renders for External Walls*, Digest 410 (October), Building Research Establishment

BRE (1996) *Assessing External Rendering for Replacement or Repair*, Good Building Guide 23, Building Research Establishment.

Hinks, A.J. and Cook, G.K. (1989) *Defects with External Cementitious Rendering*. Building & Maintenance, No. 6.

National Building Agency (1983) *Common Building Defects*, Construction Press.

# 6.7 Internal joinery

## Learning objectives

- You should be able to understand the cause and effect of a range of defects associated with items of internal joinery.
- This will establish the influence of normal use and dimensional changes of the timber.
- The problems which can occur with internal partitions will be examined in a way which allows you to identify the influence of structural movement.

## Doors

Timber doors can become distorted particularly where they were poorly hung initially. Where the frame has retained the original shape gaps may appear around the distorted door. The closing stiles of panelled doors can drop and mark the floor finish. The wider the door the greater the distorting load.

Distortion may affect the operation of any door furniture causing a lack of security or privacy. Worn or defective hinges and poor fixing to the frame can also cause doors to operate inadequately.

The traditional fire check door relied on a minimum 25 mm rebate in the frame; any distortion may compromise the ability of the door to resist smoke and fire penetration. The more modern lightweight fire check doors are commonly fitted into a frame containing an intumescent strip. Where they are fitted into a timber frame they may not perform a fire check function.

Flush doors faced with lightweight material have a poor resistance to impact damage. Since flush doors have no obvious markings to identify which way they should be hung, it is possible to fit furniture incorrectly to flush doors.

## Stairs

Older stairs may creak. This may be due to the seasonal variations in the moisture content of the timber or movement in the joint between tread and riser. Where triangular fixing blocks have been omitted or poorly fixed there

*The Technology of Building Defects.* Dr John Hinks and Dr Geoff Cook.
Published in 1997 by E & FN Spon, 2–6 Boundary Row, London SE1 6HN, UK. ISBN 0 419 19770 2

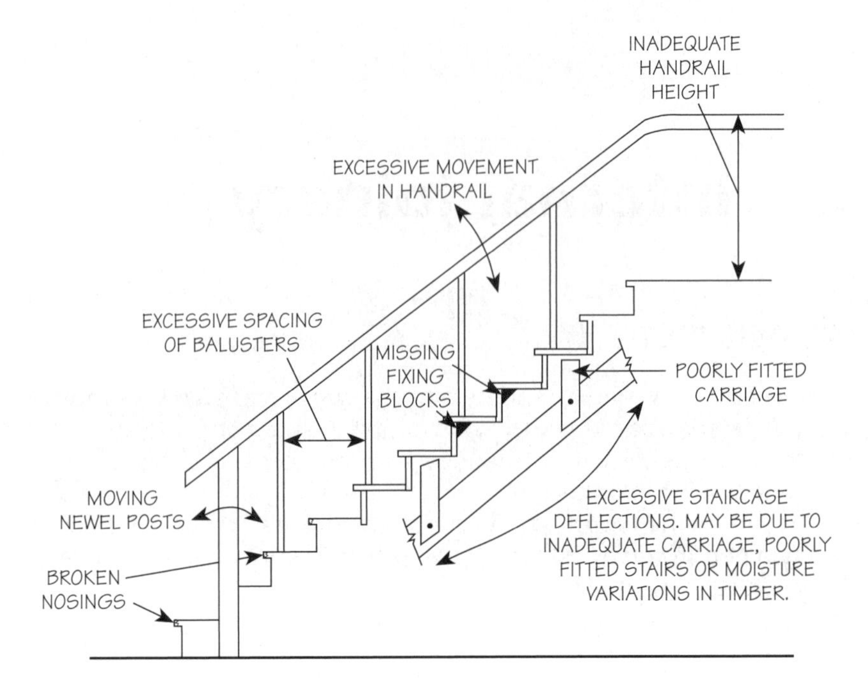


**Fig. 6.16.** General defects of timber stairs.

may be movement of the tread against the riser. This can result in cracking of the treads.

The whole flight of stairs may distort under normal load, causing wholesale differential movement of the stair components. The fixings between the carriage and the treads may be defective. Distortion of the tread under load may also cause movement in the joint between tread and string.

Treads can wear and may become permanently distorted. The nosings are particularly vulnerable.

The handrail fixing may be inadequate to support a falling adult. There may be excessive movement of the handrail, particularly when the distance between newels or other fixings is excessive. Older staircases may have generally lower handrail heights and wider baluster spacing than those of contempory design. Newel posts can become loose.

Stairs to cellars or basements which are damp or have been flooded may be rotten and collapse when used.

## Partitions

Timber partitions may be loadbearing or non-loadbearing. It may not be a straightforward process to determine the role of any particular partition. This is particularly true for timber-framed buildings. The structure should be assessed before any partitions are removed or openings cut through them. There may be considerable structural implications associated with the removal of a loadbearing partition.

Fixtures and fittings to non-loadbearing partitions commonly require noggins. These are commonly hidden behind finishing materials making location problematic, particularly where they have not been installed.

Various proprietary fixings are available although these may fail where the manufacturers' instructions have not been followed.

Ground floor timber partitions may be distorted owing to the settlement of fill or the expansion of shales under the concrete ground floor slab. Distortion can also occur where suspended floors deflect under load. Shrinkable aggregates in concrete floors can cause this deflection.

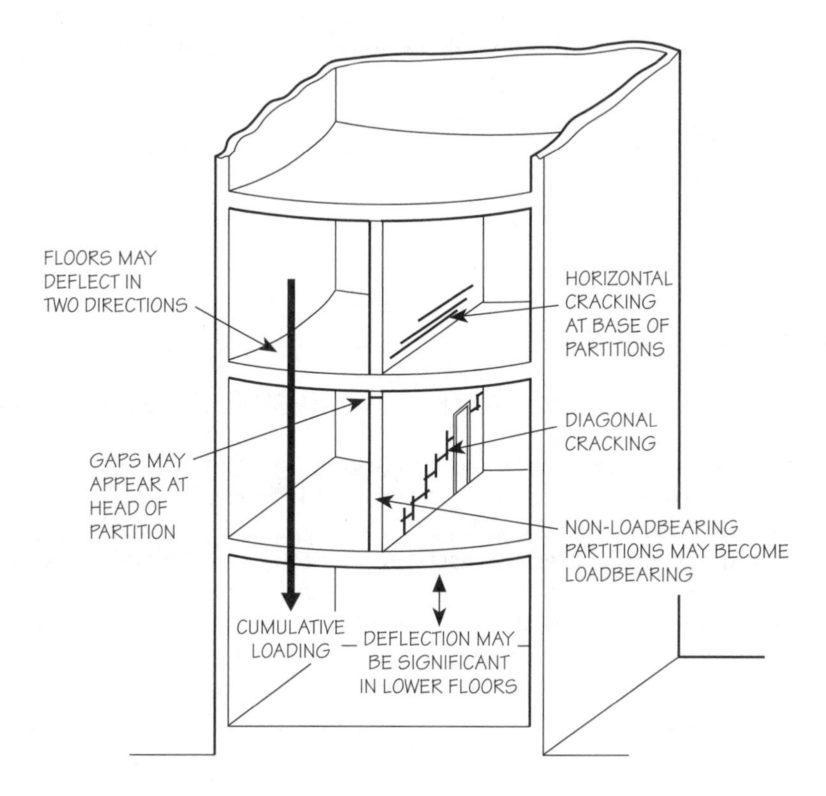


**Fig. 6.17.** Possible failure mechanisms of partitions.

Non-loadbearing partitions in upper storeys may rest on timber floor joists. Since the partitions may not be directly over the ground floor partitions, deflection of the floor can occur. This may create twisting, leaning and deflection of the partition walls.

The deflection of floors or beams supporting timber partitions may cause the cracking of finishes applied to the partition. Tapered cracking may be seen in the lower sides of the partition as it bends as tensile forces are produced at floor level. These may extend approximately halfway up the partition. There may also be horizontal cracking in the partition approximately one course above floor level. Cracking may be greater at lower floors owing to any cumulative effects of floor deflection.

If the supporting floor deflects transversely to the line of the partition there may be wholesale downward movement, which could produce a uniform horizontal opening at the top of the partition and vertical cracking where it abuts adjoining stable walls. In extreme cases a tensile crack may appear on the underside of the floor in line with the partition. Any cracks or openings in partitions may reduce their smoke and fire resistance.

## Revision notes

- Timber doors can become distorted by wear and tear and dimensional changes of the timber. This distortion may affect door furniture and rebate depth, which can be critical for fire check doors.
- Stairs can creak owing to dimensional changes caused by changes in the moisture content of the timber and/or inadequate fixings and joints in the construction. Older stairs may be unsafe.
- Do not remove partitions until a structural assessment has been made. Partitions can be distorted by floors deflecting or bowing. Any cracking of finishes may indicate the pattern of loading.

## ■ Discussion topics

- Appraise the deterioration mechanisms of timber internal doors and timber windows.
- A timber stair 1.5 m wide has a half landing and a total rise of 3.1 m. It is creaking when used. Produce detailed sketches to explain the possible causes of the creaking.
- Describe a procedure for assessing whether a partition in a two-storey dwelling is loadbearing or non-loadbearing.
- Discuss the influence of structural deformation of a building on the integrity of any internal partitions. Identify key aspects which could be used to assess the nature of any structural deformation.

## Further reading

BRE (1983) *Wood Windows: Arresting Decay (Design)*, Defect Action Sheet 13, Building Research Establishment, HMSO.

BRE (1984) *Stairways: Safety of Users – Installation (Site)*, Defect Action Sheet 54, Building Research Establishment, HMSO.

BRE (1986) *Gluing Wood Successfully*, Digest 314, Building Research Establishment, HMSO.

BRE (1987) *Fire Doors*, Digest 320, Building Research Establishment, HMSO.

BRE (1988) *House Inspection for Dampness: A First Step to Remedial Treatment for Wood Rot*, Information Paper 19/88, Building Research Establishment, HMSO.

BRE (1995) *Removing Internal Loadbearing Walls in Older Dwellings*, Good Building Guide 20, Building Research Establishment, HMSO.

BRE (1995) *Timber for Joinery*, Digest 407, Building Research Establishment, HMSO.

Cook, G.K. and Hinks, A.J. (1992) *Appraising Building Defects: Perspectives on Stability and Hygrothermal Performance*, Longman Scientific & Technical, London.

PSA (1989) *Defects in Buildings*, HMSO.

Ransom, W.H. (1981) *Building Failures: Diagnosis and Avoidance*, 2nd edn, E. & F.N. Spon, London.

CHAPTER 7

# Defects in external joinery

# 7.1 Timber window and door frames

**Learning objectives**

You should understand the nature of:

- failure within the frames due to material specification and inherent durability;
- workmanship and detailing associated with installation;
- defects/deterioration in service;
- distinctions between frame defects and defects at the interface of the frame and wall.

Rot in timber windows (and doors) is one of the most common defects cited in surveys. Problems can arise owing to the poor durability of the timber, which may be compounded by inadequate preparation and/or on-site protection during storage and installation. Poor treatment after installation, including poor painting, can also be a source of early failure problems. Secondary problems arise where unprotected timber is subject to excessive moisture movement. Poor paintwork condition could be the direct result of inadequate protection of the frames during storage and installation, leading to wet wood which does not accept paint properly. Similar localized problems can occur with damage to the priming coating due to physical damage or weathering. Unprotected/untreated ends are especially vulnerable to rot because of their propensity to take up water.

## Failures in timber and door frames generally

The traditional timbers for windows (and door frames) are softwoods. These imported timbers have been used for complete windows, although their inherent durability problems mean that windows in exposed situations and vulnerable areas generally incorporate hardwoods.

Timbers only perform satisfactorily in a window when the unit complies with necessary workmanship and material standards. Recent studies have identified that there is an increase in the number of opening lights showing signs of decay. It would seem, therefore, that established construction technology is not, or has not been, applied.

*The Technology of Building Defects.* Dr John Hinks and Dr Geoff Cook.
Published in 1997 by E & FN Spon, 2–6 Boundary Row, London SE1 6HN, UK. ISBN 0 419 19770 2

Since many failures of these windows relate to decay, the role of the protective covering and paints cannot be ignored. But it is quality workmanship, in the areas of weathering, sills, jambs and heads, which avoids permanent defects. Windows and doors need to perform as well as the surrounding wall. However, the reduction in size of timber sections over the years has meant that the critical interfaces between casement and frame have become smaller.

There are a number of typical defects associated with timber window and door frames (Table 7.1).

**Table 7.1** Common defects associated with timber window and door frames

| Defect | Cause |
|---|---|
| Decay of timber | High moisture content |
| Dry rot | May be part of an attack from elsewhere |
| Wet rot | Most common form of attack |
| Putty failure | Exposure and/or poor protection<br>Lack of slope to weathered surfaces<br>Condensation<br>Applied to moist timber |
| Wet timber | Protective film failure<br>Opening joints in frames and casements |
| Decayed sills | Lack of water bar between stone and sub-sill<br>Inadequate fall to top surface<br>No throating |

## Weatherstripping

Grooves are commonly inserted to channel away water that breaches the external joint. In certain cases these grooves have assisted in allowing water direct entry.

Although there are many instances where wet timber has not become defective, where this condition persists for extended periods the risk of attack is high. The need for thorough inspections and sensible maintenance regimes should ensure that these faults are rectified before they produce defective windows.

Certain dowelled joints can be a source of decay, either through the use of inferior timber for the dowel, or through insufficient adhesive to seal the joint around it. This may weaken corners of the casement and frame and add to the problems of distortion.

## Casement window frame defects

Owing to moisture and general use, frame distortion is possible. Where casements fit badly the risk of impact damage increases as occupants close and open the window. This can be caused by paint runs on the frame edges forming blobs and runs. Paint runs also prevent adequate sealing of the meeting surfaces.

Thick timber sections are more likely to suffer from resin seepage, although the relatively small sections commonly used can also suffer from this problem. Knotting compounds which seal these surfaces are frequently omitted.

Joints between the timber frame and opening construction can be defective. The setting-in of the frame must not allow water access to the back of the frame, although the provision of priming to concealed surfaces offers some protection.

Problems with rot are most likely in fixed and opening frames at the joints between the bottom rail and the stiles. Frames which have opening lights

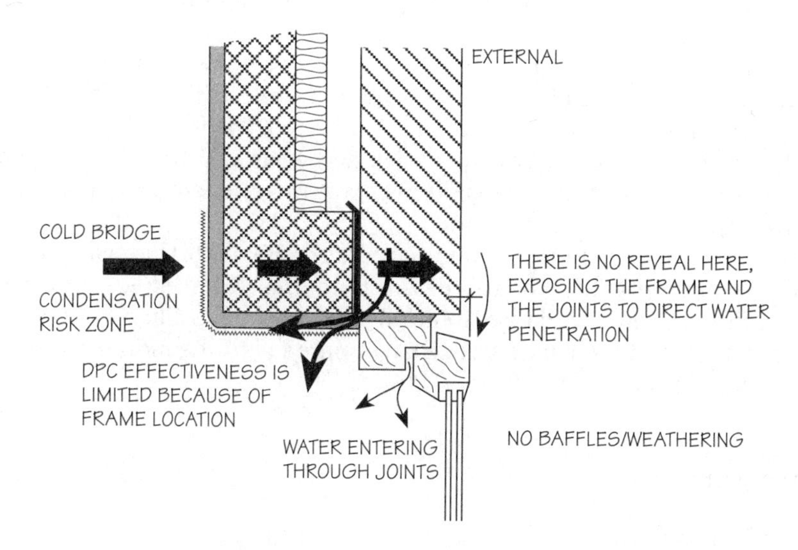


**Fig. 7.1.** Poor design and/or construction detailing at head of window.

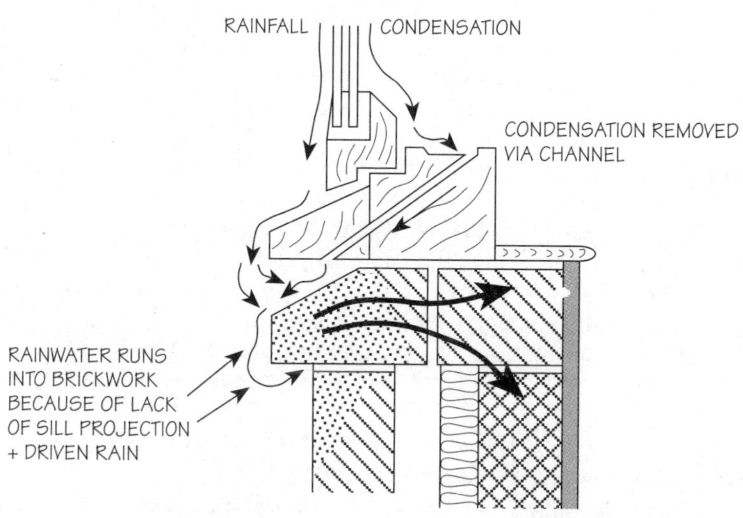


**Fig. 7.2.** Inadequacy in sill projection/ weathering leading to wall dampness.

are most susceptible because of the easing of joints which occurs here. Faults may also occur at the connections between mullion and cill; in both cases these are the common locations for water collection, and consequent rot. Early detection may allow localized repair. Susceptibility to rot is increased where the preservative treatment of the timber is of poor quality, and where inadequate weathering slopes and a lack of condensate drainage grooves allows the collection of water. Problems also arise at the joints between frame and wall, where water may penetrate behind poorly sealed joints.

## Sliding sash windows

Sliding sash windows make a significant visual contribution to the elevations of many fine buildings, and where it is sensible, these are normally retained during any repair and refurbishment work. The catches to these windows are often ineffective, and many are covered with many layers of paint. Replacement glass can be heavier than the original and unbalance the sash weights. Sash cords can fail owing to age and wear and should be replaced with good-quality cord or corrosion-resistant metal chain; it is also appropriate to replace the cords in pairs.

There are a large number of weathered surfaces with this window type, and all the puttied joints are sources of potential water egress. The movement of the sashes can cause wear to the protective film, although the increased depths of the window reveal compared with the modern casement frame may offer some protection.

### Revision notes

- Rot is the most common problem with frames.
- Causes lie in the material specification and preparation, installation, protection and use. Failure may be due to a combination of these factors.
- Specification and preparation problems include treatment and priming, which may be affected by storage and handling, also pre-installation trimming.
- Installation problems arise from poor design and/or site detailing, both of which are critically dependent on workmanship.
- In use, poor design detailing of jamb/wall interfaces, weathering and drainage provisions can accelerate deformation, particularly at vulnerable joints in frames and opening casements. Water collection is a common agent with such failures.
- Secondary problems of moisture entry also occur, frequently related to excessive moisture-induced movement coupled with weakened joints.
- Problems can also arise with weak frames carrying heavy dead and induced loads of double-glazed units.
- Jointing of glazing is also a source of water ingress and requires attentive maintenance.
- Other problems arise from resin seepage at joints.

## ■ Discussion topics

- Compare and contrast the common defects which occur with timber window and door frames. Draw out the similarities and the distinctions between in-service failure mechanisms.
- Discuss the causes of defects by categorizing them according to design, installation and service. Identify the interactions between defect mechanisms in these three categories.
- Produce a survey checklist for assessing external joinery, and appraise it in use.

## Further reading

Addleson, L. (1992) Window openings, Study 16 (Failures in context), pp. 85–9, in *Building Failures: Guide to Diagnosis, Remedy, and Prevention*, Butterworth-Heinemann, pp. 85–39.

BRE (1979) *Preservative Treatments for Softwood Joinery Timber*, PRL, Technical Note 24 (revised), Building Research Establishment.

BRE (1982) *Wood Windows and Door Frames: Care on Site during Storage and Installation*. Defect Action Sheet DAS 11 (December), Building Research Establishment.

BRE (1983) *Wood Windows: Resisting Rain Penetration at Perimeter Joints*. Defect Action Sheet DAS 15, Building Research Establishment.

BRE (1983) *Wood Windows: Preventing Decay*. Defect Action Sheet DAS 14 (January), Building Research Establishment.

BRE (1992) *Selecting Windows by Performance*, Digest 377, Building Research Establishment.

Carey, J.K. (1980) *Avoiding Joinery Decay by Design*. Building Research Establishment Information Paper 10/81 (July).

Carruthers, J.F.S. and Bedding, D.C. (1981) *Weatherstripping of Doors and Windows*, Building Research Establishment Information Paper IP 16/81.

Edwards, M.J. (1983) *Window to Wall Jointing*. Building Research Establishment Information Paper IP 7/83.

Hollis, M. (1982) Windows – Type and inspection techniques. *Structural Survey*, **1**, (3), 269.

# 7.2 Glazing problems

### Learning objectives

- You should understand the specific problems related to glazing in windows, many of which are irrespective of the frame materials.

Failures in double-glazed units tend to occur at the edge seals, which may become defective in use. This may be the result of defective manufacture or where exposure to sunlight and prolonged damp breaks down the seal. The ultraviolet radiation is particularly damaging to polysulphide and polymethane sealants. The seals will also be under loads due to wind load and thermal movement. The consequences may be splitting in the seal leading to water penetration and consequent localized damage to the frame; also, water may permeate the void between the double-glazing panes leading to misting and possible mould growth in the insulating space and a reduction in insulating effectiveness. Misting may occur during changes in temperature, cycles of which will produce a scum on the internal face of the outer pane.

Front-sealed glazed units are particularly vulnerable to water entrapment following any seal failure. Faulty double-glazed units require replacement since repair is not feasible.

## Cracking and devitrification of glass

The cracking of the glass in metal casements is usually a sign that corrosion is occurring. The production of rust is associated with a volume increase in the steel frame. Something has to give, in this case the glass.

Glass is a supercooled liquid. With age this material can become more and more crystalline, and therefore more brittle – this is termed devitrification. Discoloration is also occasionally evident. The devitrification of glass occurs over a long period of time, similar to that usually associated with the advanced corrosion of the frame. Devitrification may also be a factor in occurrences of cracked glazing.

*The Technology of Building Defects.* Dr John Hinks and Dr Geoff Cook.
Published in 1997 by E & FN Spon, 2–6 Boundary Row, London SE1 6HN, UK. ISBN 0 419 19770 2

## Revision notes

- Failures in modern double-glazed units may occur as a result of overloading of the frame.
- Also as a consequence of edge seal failure due to physical damage during assembly or installation, or deterioration due to prolonged wetting and/or exposure to sunlight.
- Symptoms of failure in glazed units include intermittent misting which may be accompanied by localized rot in timber frames.
- Cracking may occur if ferrous metal frames corrode and squeeze the glazing.

---

## ■ Discussion topic

- Create a series of annotated sketches indicating the path and mechanism of water entry into double-glazed units, indicating the high-risk areas and good design features that can be found in frames to protect against this problem.

---

## Further reading

BRE (1994) *Thermal Insulation: Avoiding Risks*, Building Research Establishment Report, ch 41 pp. 35–42.

Garvin, S.L. and Blois-Brooke, T.R.E. (1995) *Double Glazing Units: a BRE Guide to Improved Durability*, Building Research Establishment Report.

# 7.3 Metal window frames

**Learning objectives**

You should understand:

- the mechanisms of failure in the framing material;
- the particular types of defect that can occur in metal windows.

Steel window frames can suffer from a range of defects which can cause failure in the component when of sufficient maturity.

The most common failure of steel windows is that they corrode, and with the corrosion comes expansion and distortion. Many of the steel windows which were built into projects during the 1920s and 1930s were ungalvanized and therefore particularly prone to corrosion.

The chemical processes related to corrosion are considerable, but it is essential to the understanding of the defects associated with steel windows that an outline of the processes is given. They can be simplified into three basic types: oxidation, electrolytic corrosion and acidic corrosion. Since windows are exposed to the atmosphere they can suffer all three forms of attack.

## Oxidation

This can be disruptive, even where the oxides (the new compounds formed by the reaction of oxygen and the parent metal) formed are small. Since the oxidation process leaves gaps on the surface of the metal, the oxidation is continuous. The mechanism by which the corrosion of the iron in the steel is further accentuated involves the carbon in the steel. This can react with oxygen to form gaseous carbon dioxide which is then absorbed into the atmosphere, leaving a porous and weakened surface to the steel. Because the surface area in contact with the air has been increased, it is more likely to suffer from further corrosion. Newly exposed surfaces react to release carbon dioxide, and so on.

## Electrolytic corrosion

This describes the process which is driven by the fact that at the atomic scale, metals are made up of nuclei around which electrons are orbiting.

*The Technology of Building Defects.* Dr John Hinks and Dr Geoff Cook.
Published in 1997 by E & FN Spon, 2–6 Boundary Row, London SE1 6HN, UK. ISBN 0 419 19770 2

**Fig. 7.3.** Buckling of metal windows due to expansive effect of corrosion.

When in contact with water, the metallic atoms which are characteristic of the particular metal dissolve at different rates. As the atoms dissolve they tend to leave behind some electrons which can flow from one metal to another via the surrounding water. This transport of electrons to other metals causes the original metal to continue to dissolve. This cycle depletes the metal and reduces its strength. This is problematic where different metals are connected directly or by water, but it is also possible for differences to exist within the same metal which can cause the same process to occur. It is therefore possible for metals in solution to set up an electrolytic cell if there is variation in the metal composition. This produces rust which is then deposited on the metal surface. The deposits are commonly found close to the regions of depletion and are therefore a visual indication of corrosion attack.

## Acidic corrosion

This type of corrosion occurs when metals are in contact with rainwater, since this can be acidic. Where the electrons are held in the parent metal, they can react with the hydrogen in the acid to produce hydrogen gas. This will drive on the reaction to deplete the metal.

As with any corrosive reaction which increases the surface area of the metal, the risk of further attack is increased. Adequate protection therefore is essential if corrosion is to be prevented.

## Faults in metal windows

The thickness of frame section must be adequate to obtain the long-term durability of the window, since reduced thickness will bring with it reduced strength and increase the risk of bowing. This problem can be found with older windows produced when corrosion theory was not fully developed, and which were seriously undersized.

Particular faults associated with double-glazed window units arise with aluminium frames fabricated without a thermal break. These frames can have significant thermal bridging problems, whilst aluminium suffers specific problems with discoloration and corrosion due to atmospheric pollution.

In some instances the frame strength may be insufficient to cope with the weight of large double- or triple-glazed units, especially when deflected by gusting.

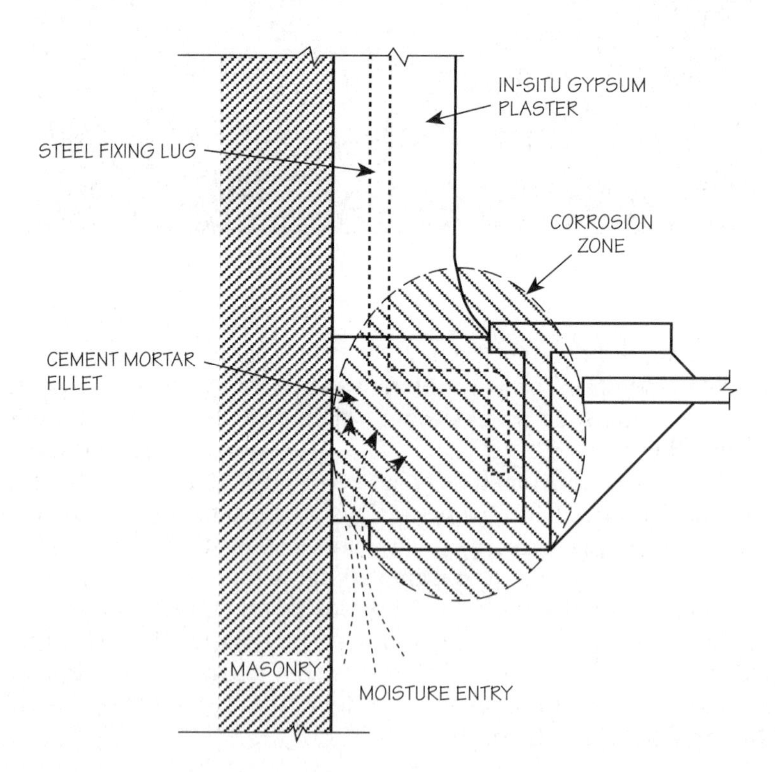


**Fig. 7.4.** Corrosion at steel window frame/wall junction.

## Common defects

The metal sections are regions of high thermal transmittance; this can create a series of cold bridges across and around the edge of the window. These in turn will increase the incidence of condensation at just the place where it should be avoided, since the condensate will drain into the back putty areas or any abutting plaster where it will lie, slowly acting upon unprotected metal and also creating a site for mould growth. Metal casement putty is essential. Where the casement putty has been applied to damp surfaces, the adhesion can be permanently lost, which will increase the number of cracks and crevices within which such corrosion and water entry can occur. Although the amount of retained water is small, this process can continue for an extended period and produce serious deterioration. Kitchens and bathrooms are at particular risk. Owing to the necessarily complex shape of the metal window section, there are usually a range of potential corrosion sites.

It is worth noting here that discontinuity cavity insulation at cavity closings around window frames can also be a cause of cold bridging faults appearing on the frame.

Excessive paint build-up can distort the frames because paint runs on leading and trailing edges amalgamate to form blobs and runs, which prevent adequate sealing of the meeting surfaces. Joints between the metal frame and the wall opening may be defective. The common details that rely on the plaster thickness and a butt joint into masonry can allow the direct entry of moisture.

**Table 7.2** Defects in metal window frames

| Defect | Cause |
|---|---|
| Cracking of glass | Rusting of frame |
| Water ingress around frame | Failure of the perimeter joint |
| Gaps around the meeting surfaces | Thermal movement<br>Uneven paint thickness<br>Air demand of appliance in the room<br>Site damage |
| Staining of surrounding masonry | Corrosion products from frame |

### Revision notes

- Most common failure in ferrous metal windows is corrosion due to oxidation, electrolytic corrosion or acidic corrosion.
- The consequent expansion caused by corrosion can cause damage to the glazing, and inoperable opening windows. Joints may be affected in severe cases.
- Metal windows in general suffer from problems of high thermal transmission, except where modern thermal break detailing is included.

- Aluminium frames may suffer from atmospheric pollution damage.
- Connections between the frame and wall, and frame and glazing are common locations for water penetration and defects due to inadequate interface detailing.

---

## ■ Discussion topic

- Describe the major sources of failure in metal windows, and the likely locations of such failure and consequences. Design the remedial options.

---

## Further reading

BRE (1990) *Assessing Traditional Housing for Rehabilitation*, BR167, Building Research Establishment, ch. 9.

BRE (1994) *Thermal Insulation: Avoiding Risks*, Building Research Establishment Report, (ch. 4, pp. 35–42).

Hollis, M. (1982) Windows – Types and Inspection Techniques. *Structural Survey.* **1** (3), 269.

# 7.4 Plastic windows and doors

## Learning objectives

- You should understand the mechanisms of degradation of uPVC and the common symptoms of failure in plastic windows and doors.
- Since repair is frequently not an economic or practicable option it is important to be able to identify deficiency, reliably.

Prior to the 1970s there was little use of plastic windows or doors, and as with all new building materials the pattern of uptake and analysis means that there is relatively little discussion in the texts on faults with plastic windows.

The potential faults can be categorized according to the material and design specification, the installation and detailing associated with this, and in-service durability. Unlike timber, it is not usually practical or cost-effective to repair sections of uPVC windows and hence comparatively minor damage will usually mean that the entire window requires replacing.

## Structure

Reinforcement is quite critical in uPVC windows since plastic is not as stiff as wood and metal and the frames require some internal support against the often considerable loads from the glazing. Distortion is a distinct possibility in under-reinforced sections. It is also important that the window frame is not installed in a twisted or racked manner. Fixing is also important, especially in upper windows and exposed locations. Coupled with potential problems of brittle fracture during manufacture, installation or service in cold conditions can lead to potential durability problems.

*The Technology of Building Defects.* Dr John Hinks and Dr Geoff Cook.
Published in 1997 by E & FN Spon, 2–6 Boundary Row, London SE1 6HN, UK. ISBN 0 419 19770 2

## Reversible and irreversible shrinkage

The thermal expansion coefficient of plastics, particularly the emerging coloured sections, is higher than that of traditional window framing materials, and the scale of reversible thermal movement that has to be accommodated can be significant in large windows in locations exposed to thermal gain. In some circumstances the thermal efficiency of the window can lead to differential expansion across its thickness, producing bowing as the external surfaces reach greater temperatures than the internal surfaces. This can also lead to problems with heat reversion which produces an irreversible shrinkage.

## Stability

The earliest unplasticized PVC window sections suffered from embrittlement upon prolonged exposure to sunlight, and the modern uPVC versions use a variety of chemically stabilized plastics in their construction. Modern uPVC windows and doors are now available in a range of colours.

Under prolonged exposure to sunlight, rain and pollution the uPVC surface degrades and may become chalky and dull. The impact resistance may also decrease.

As with curtain walling and other impervious façades, water run-off can place great loads on the sealants, and the drained and vented systems appear to perform better at discharging water run-off and limiting the build-up of water within the frame sections.

Research suggests that the majority of defects tend to be due to poor workmanship on installation rather than in manufacture. Windows using uPVC may of course suffer the same glazing defects as other frame materials. Some front-beaded systems have security problems, however.

Characteristic potential sources of defects occur at the sealing of the window frame to the structural joining, in terms of water penetration; and in cases where the cavity insulation is stopped short, of cold bridging in the surrounding wall.

### Revision notes

- A range of potential problems occur with plastic windows, mostly due to installation workmanship rather than manufacture.
- Design: strength of section, position and thermal effects of reinforcement, beading detailing.
- Installation: racking, twisting, beading and fixing detailing.
- In service: thermal movement and irreversible shrinkage, especially with dark-coloured sections. Shape stability problems, creep, weatherstripping failure and embrittlement. Failure/detachment of decorative foil.

## ■ Discussion topics

- Compare and contrast the defects occuring in metal, timber and plastic window frames, and assess the relative weaknesses of each frame type.
- Produce a checklist for defects in uPVC windows, and attempt to rank them in terms of (a) likely occurrence and (b) seriousness of the consequent defect.

## Further reading

BRE (1995) *PVC-U Windows*, Digest 404 (April), Building Research Establishment.
BRE (1994) *The Role of Windows in Domestic Burglary*, Building Research Establishment Information Paper IP18/94.

PART 3

# Elements

CHAPTER 8

# Defects in elements: soil and foundation-related problems

# 8.1 Subsidence

## Learning objectives

You should be able to:

- distinguish between the causes and symptoms of various forms of subsidence;
- characterize reversible and irreversible soil movements.

*Subsidence* is a downward movement under applied load. It can occur as a wholesale movement whereby the building progressively settles with the soil. This may be small, in the order of millimetres. According to the soil type, it may manifest itself in the first few years following construction, or almost at once (e.g. in granular soil). Clays may react over a long period as they squeeze out the pore water gradually. Minimal subsidence occurring uniformly may produce little in the form of structural problems for the building. It is more common for partial or differential subsidence to occur, however, and it is this form of movement which is generally more destructive, especially to building services. This usually results in differential movement in the foundations, causing profound disruption to the structure and fabric of the building. Mining subsidence is usually predictable (at a gross scale) from National Coal Board records. Special foundation/structure forms have been devised for such problems (where foreseen), such as raft foundations and the CLASP system.

The cause of subsidence may be as straightforward as poor bearing capacity or soil erosion. It is distinguishable from settlement, which is a movement within a structure due to a redistribution of loading *within* the structure (or soil). The scale of settlement is variable but can be considerable and dramatically affect the viability of the DPC as well as services and access.

## Mining subsidence

In instances of mining subsidence the effect may be very sudden, as with seismic activity, as the supporting soil strata re-establish their equilibrium by subsiding under the force of gravity. A range of sophisticated measurement techniques are now available to allow continuous monitoring of the

*The Technology of Building Defects.* Dr John Hinks and Dr Geoff Cook.
Published in 1997 by E & FN Spon, 2–6 Boundary Row, London SE1 6HN, UK. ISBN 0 419 19770 2

ground movement associated with gradual subsidence caused by worked-out mine workings.

The effect of mining subsidence is in part determined by the depth and position of the mine workings, the seam thickness, any partial storage (replacement of excavated material) and the width/depth ratio; also the time since working out. Clearly the nature of the soil is also significant in determining the local response to mining subsidence – in particular this relates to the type and distribution of the soil and the rock disposition and fault locations, factors which should be studied carefully. Hence, local knowledge will be essential here. Failure is likely to occur directly above seams, and the locality of the effects is defined by the depth of the workings – the width of subsidence is usually approximately half the depth of the workings from the surface.

The subsidence proceeds as a wave along the direction of the workings, damaging the buildings by tension as the wave starts to pass under them (accompanied by a slight rise in level), then by compression as it progresses, completes its pass and leaves a relatively uniform (but subsided) level. The local symptoms include depressions in the ground. The type of building structure will affect its response to the mining subsidence. Important factors include the size, shape and type of superstructure, the foundation forms and age/ductility of materials. Resilient and flexible structures may prove to be structurally safe although distorted upon testing. In such cases the limit of immediate problems may be cracked plaster and the sticking of doors and opening windows. Commonly, the underground services to the building will be severed, however.

Subsidence symptoms are varied. There is usually cracking, especially with rapid changes in the soil characteristics. Cracks may be quite wide (25 mm). Older, soft buildings may accommodate some distortion, the symptoms here including out-of-plumb walls rather than overt cracking. The condition of the cracks can provide a useful indication of the age of the fault.

In affected buildings the foundations and walls or structural frame may exhibit distortion and/or tensile cracking, which usually reduces in width with height. The compression-related symptoms may include overlapping of roof tiles, bulging of walls and some horizontal cracking of brickwork. Ground cracks of up to 150 mm have been observed. Structural failure is therefore a possible consequence as tensile and compressive ground strains create curvature in the soil. This is where the maximum damage occurs, and in cases of curvature there is the likelihood of differential subsidence.

It is also common for cracking to pass through the DPC, thus assisting in distinguishing between soil/foundation movement and thermal or moisture movement (which rarely bridge the DPC). Internal walls may also exhibit cracking, perhaps in a complex manner.

Where differential settlement occurs, the structure may tilt or fracture. Since buildings usually fail in tension or twisting, the cracking patterns may assist in identifying the location(s) of settlement (and/or heave conversely). However, since the failure in the building is a composite of its own structural nature and orientation with the advancing subsidence face (which will also vary in intensity), the cracking may be of a complex nature. The structural integrity of the building should be assessed.

**Fig. 8.1.** Mining subsidence can affect buildings and their general vicinity. In this view the door pillars are vertical.

## Revision notes

- A downward movement, whereby the building progressively settles with the soil.
- May be rapid or prolonged.
- Usually small-scale. Mining subsidence can be on significant scale however, and produces complications because of curvature of the soil during adjustment.
- Symptoms depend on the nature of the structure – factors include ductility, strength and shape with respect to the pattern of soil movement.
- Irreversible subsidence occurs with mining subsidence and erosion.
- Reversible subsidence may occur with clay soils as they desiccate and rehydrate cyclically.

## ■ Discussion topics

- Explain the differences between subsidence and settlement, and comment on the distinguishing forms of defect that result.
- Describe the range of factors affecting the appearance of symptoms of mining subsidence.

## Further reading

Bell, F.G. (ed.) (1975) *Site Investigations in Areas of Mining Subsidence*, Newnes Butterworth, London.

Bickerdike, Allen, Rich & Partner and O'Brien, T. (1971/72) House built in a wood, in *Design Failures in Buildings*. 1st series, Building Failure, Sheet 7, George Godwin.

BRE (1991) *Why Do Buildings Crack*? Digest 361 (May), Building Research Establishment.

Cook, G.K. and Hinks, A.J. (1992) *Appraising Building Defects: Perspectives on Stability and Hygrothermal Performance*, Longman Scientific & Technical, London, p. 136.

ISE (1994) *Subsidence of Low Rise Buildings*, Institution of Structural Engineers (March), pp. 22–29, 82–87.

Longworth, T.I. (1988) *Techniques for Monitoring Ground Movement Above Abandoned Limestone Mines*, Building Research Establishment Information Paper IP 1/88 (January).

Shabka, G. and Kuhwald, K. (1995) Subsidence and the associated problems with reference to low-rise housing. *Structural Survey* **13** (3), pp. 28–35.

# 8.2 Clay soils

## Learning objectives

- You should be able to explain how the dessication/rehydration cycle operates in clays, and how this affects buildings.
- You should also be aware of the role of other factors in clay movement faults, such as trees, substantial vegetation, and the building form and foundation style.

The volumetric and load bearing sensitivity of clay soils to moisture content produces a range of defects in many areas of the UK. The stiff plastic clays common to the south of England tend to undergo the greatest volumetric change and hence these defects are more significant (vertical reversible, seasonal movements of 25 mm are possible). Inherent clay soil problems may be exacerbated by the presence of trees, drought conditions and flooding (or burst pipes). Clay soil movement caused by reversible water content changes will be cyclic.

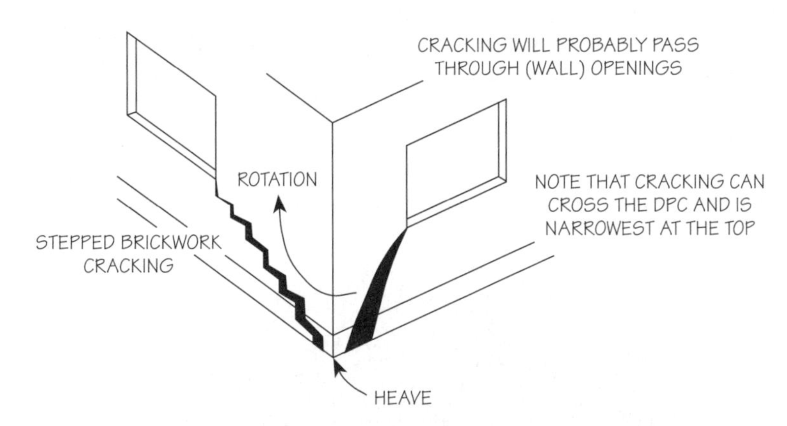


**Fig. 8.2.** Clay heave cracking symptoms at the corner of a building.

Clay soil will exhibit heave with increases in moisture content (rehydration), which can cause lifting at the corners of buildings. Heave is possible where established trees are removed, and the water-table rises as it adjusts to the reduced demand for water from the surrounding vegetation. This can be particularly complicated to diagnose or predict. The centre of the building may also be disrupted in drought conditions as the central plug of

*The Technology of Building Defects.* Dr John Hinks and Dr Geoff Cook.
Published in 1997 by E & FN Spon, 2–6 Boundary Row, London SE1 6HN, UK. ISBN 0 419 19770 2

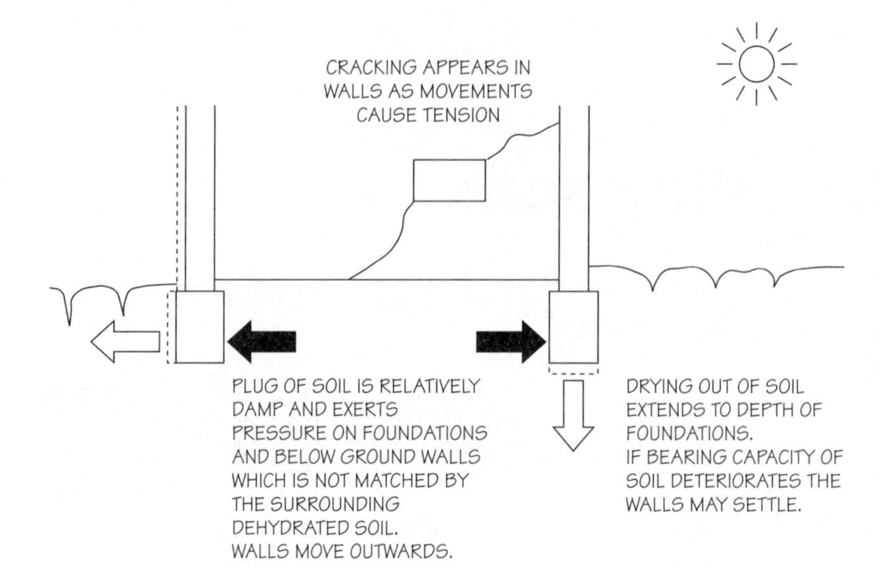


**Fig. 8.3.** Movement of clay soil under buildings.

soil protected by the building retains moisture whilst surrounding soil dries out. This is termed 'hogging'. Desiccation is a common cause of subsidence damage to buildings.

The symptoms of central hogging include vertical (tensile) cracking of the building, which increases in width across the height of the building. Such cracks usually bridge the DPC, and may be accompanied by spreading of roof coverings under tension also. The cracking may be duplicated in the opposing elevation and extend across floors and also affect internal walls.

Centralized drying out, perhaps due to tree roots or the presence of substantial vegetation, will produce reciprocal symptoms as the lower section of the building spans across a void caused by the shrinking support soil, placing the foundations and lower walls in tension. This is sagging failure. Cracking symptoms in such cases will be wider at the base, and will pass through the DPC. Similar symptoms may occur with localized weak points in other soils. In instances of desiccation of clay soils it is common

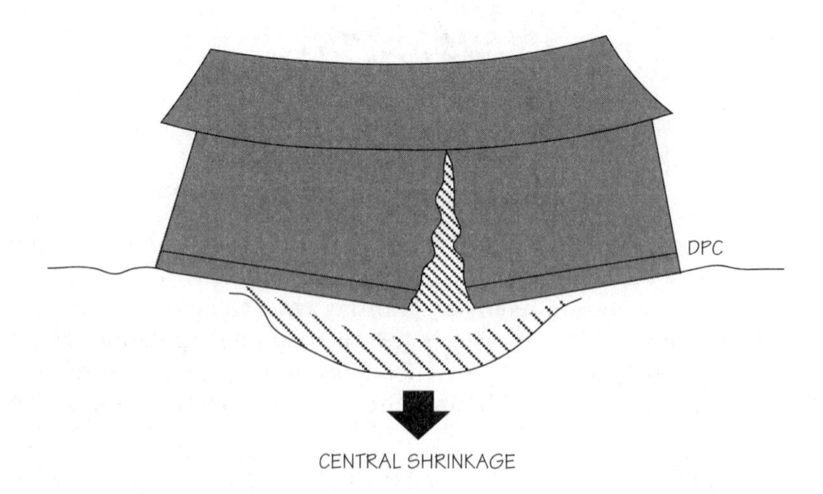


**Fig. 8.4.** Drying-out of central clay soil produces tensile cracking in the structure and substructure as it is placed in tension.

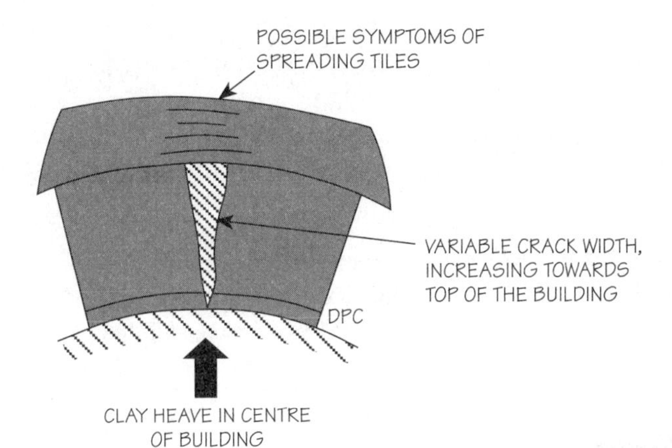


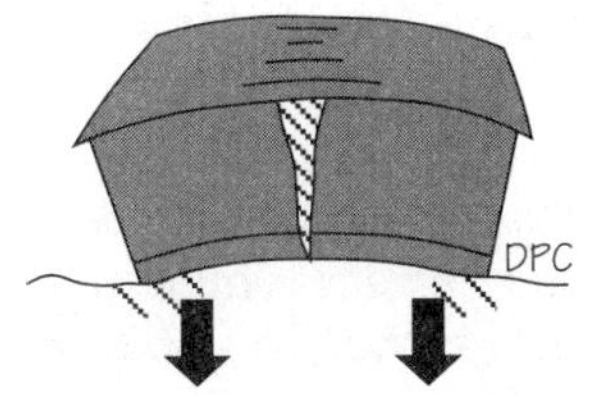


**Fig. 8.5.** The symptom similarity between clay heave and shrinkage. In both cases the severity of cracking may be seasonally variable. Monitoring may be appropriate.

for the soil to appear dry and friable, with cracking at the surface (which can be of dramatic proportions).

Cracking may also occur in localized areas of the building, such as at corners.

Clay shrinkage cycles will be accentuated by the presence of trees. Converse symptoms may occur at corners with heave. The perimeters of buildings are more susceptible to changes in water content than the centre, therefore edge/corner faults are more commonplace. As with all cracks, the fault will take the line of least resistance and is therefore likely to pass through openings.

## Revision notes

- Reversible movement occurs in clay soils as they desiccate (dry out) and rehydrate cyclically. Cracks tend first to appear after long periods of dry weather.
- Desiccation produces subsidence which usually affects the perimeter of the building. Differential settlement may occur across the plan of the building, including hogging as the plug of soil under the centre of the building is protected from drying out and therefore does not shrink correspondingly.
- Cracking is cyclic according to season and rehydration of clay.

- Trees or substantial vegetation exacerbate the desiccation problem, locally.
- Rehydration produces swelling of the clay as it reverts to its nominal size or swells beyond it. Heave occurs which disrupts the building.
- Buildings may exhibit cyclic opening and closing of cracks. Cracks will usually be open in summer and closed in winter.
- Cracks in south-facing walls or near trees tend to be largest.

---

## ■ Discussion topics

- Distinguish between the symptoms of building failure under the rehydration or desiccation of clay soils.
- What effect would the season have on the nature of clay-induced failures that the building presented, and what steps would you take to clarify the role of clay in any defects?
- How would you distinguish cracking due to clay soil movement from other building cracking?

---

## Further reading

BRE (1991). *Why do Buildings Crack?* Digest 361 (May), Building Research Establishment.

BRE (1993) *Low Rise Buildings on Shrinkable Clay Soils*, Digests 240, 241, 242, Building Research Establishment.

Cheney, J.E. (1988) 25 years heave of a building constructed on clay, after tree removal. *Ground Engineering*, **21**(5), 13–27.

ISE (1994) *Subsidence of Low Rise Buildings*, Institution of Structural Engineers (March).

Richardson, C. (1996) When the earth moves, *Architect's Journal*, 22 February, 40–42.

Richardson, C. (1996) Treating ground movement. *Architect's Journal*, 22 February, 45–47.

Shabka, G. and Kuhwald, K. (1995) Subsidence and the associated problems with reference to low-rise housing. *Structural Survey* **13**(3), 28–35.

# 8.3 Trees and buildings

## Learning objectives

You should understand the key mechanisms of tree-effect on buildings, including:

- direct disruption by root-induced upheaval;
- the more serious problem of root drainage of moisture (especially in clay soils).

The proximity of trees (or other substantial vegetation) to buildings may cause significant soil shrinkage due to drying. This is a seasonal effect usually, and is most dramatic in clay soils. The root radii of trees vary and can be especially significant for poplar, willow and oak. It is common for tree-related foundation defects to occur within a radius similar to the height of the tree (or less). This may increase to 1.5 times the radius for groups of trees, with mature heights of 20–30 m in ideal conditions for trees such as oak, sycamore, lime and beech trees. This is reduced when grown in heavy clay.

Coniferous trees are generally agreed to extract less water from soils than deciduous, although this does not extend to their risk to buildings being considered minimal. It is commonly assumed that mature trees will often only severely affect the soil in extreme conditions, and it is their removal which may trigger serious problems.

It is important to appreciate the phenomenon of soil recovery, which occurs after trees have been removed. The removal of water will cease and the water-table level will increase locally, producing heave failure. Clearly this type of causal factor may be difficult for the surveyor to identify if there is no residual evidence of the trees. This effect can also occur over a long period (25 year movement has been recorded).

Trees may also interfere with drains, the cracking of which and subsequent soil washing by waste water can lead to localized pockets of soil failure, and will encourage root growth in that area.

Tree roots can also exhibit direct disruptive capacity, and lead to damage in walls. This is usually problematic in lightweight structures – the forces involved are relatively small.

*The Technology of Building Defects.* Dr John Hinks and Dr Geoff Cook.
Published in 1997 by E & FN Spon, 2–6 Boundary Row, London SE1 6HN, UK. ISBN 0 419 19770 2

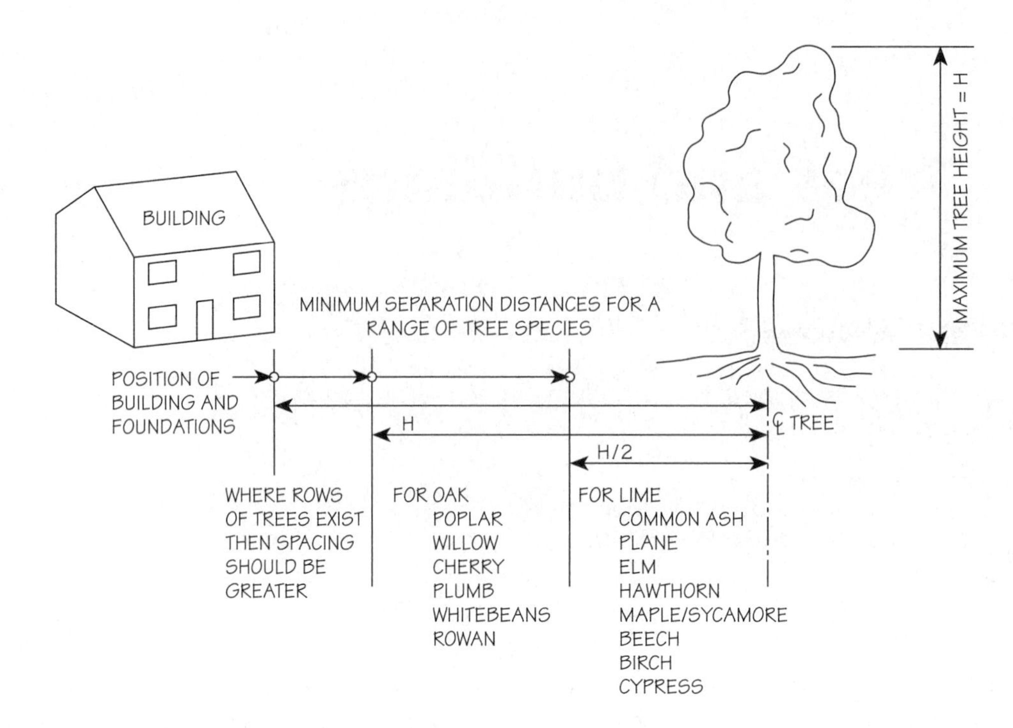


**Fig. 8.6.** Influence of tree species on separation distance.

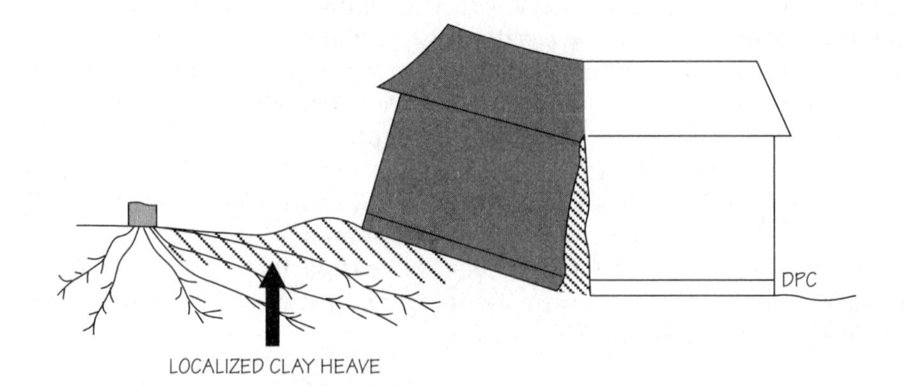


**Fig. 8.7.** Expansion at corner of building causes lifting. This may be caused by removal of tree(s) leading to localized increase in water-table level.

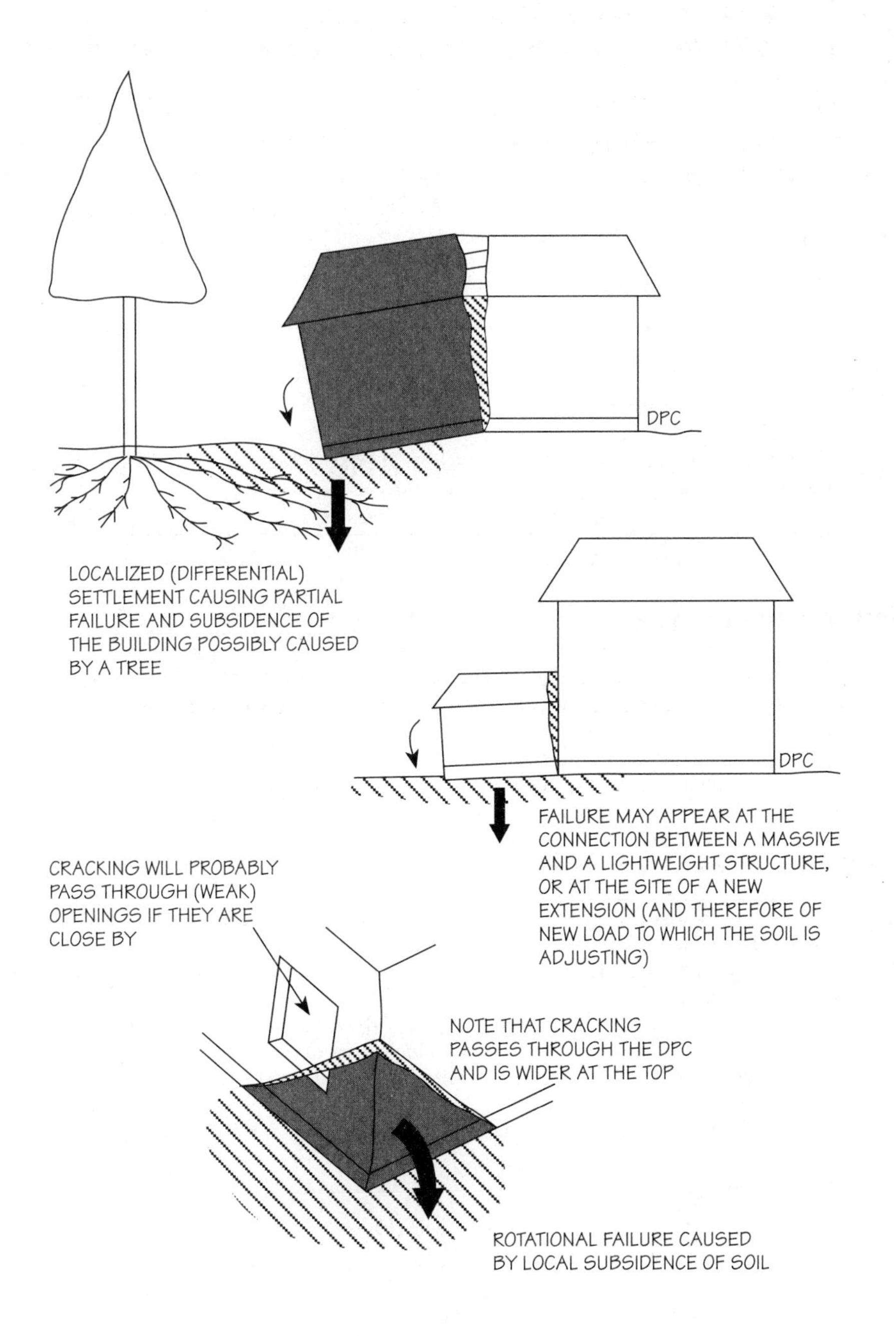


**Fig. 8.8.** Localized (partial) subsidence and settlement cracking in domestic buildings.

## Revision notes

- A seasonal cyclic effect, most marked in clay soils.
- Deciduous trees considered to be most significant. Coniferous less significant, but not insignificant.
- Problems with existing mature trees, new trees maturing, and removal of trees – symptoms may differ.
- Effects of removal can be prolonged.
- Secondary defects (e.g. drain damage) can cause their own problems.

## Discussion topic

- Explain the mechanism by which trees produce building defects. Consider the remedies solutions.

## Further reading

Aldous, T. (ed.) (1979) *Trees and Buildings: Complement or Conflict?* CIRIA Technical Note 107.

Aldous, T. (1979) *Trees and Buildings: Complement or Conflict*, RIBA Publications and the Tree Council.

BRE (1981) *Assessment of Damage in Low-rise Buildings, with Particular Reference to Progressive Foundation Movement*, Digest 251, Building Research Establishment, HMSO.

BRE (1987) *Foundations on Shrinkable Clay – Avoiding Damage Due to Trees*. Defect Action Sheet DAS 96, Building Research Establishment.

BRE (1987) *Influence of Trees on House Foundations in Clay Soils*, Digest 298, Building Research Establishment.

Cheney, J.E. (1988) 25 years heave of a building constructed on clay, after tree removal. *Ground Engineering*, **21**(5), 13–27.

Cutler, D.F, and Richardson, I.B.K. (1991) *Tree Roots and Buildings*, 2nd edn, Longman Scientific & Technical, London.

Desch, H.E. and Desch, S. (1970) *Structural Surveying*, Griffin.

Hill, W.F. (1982) Tree root damage to low rise buildings. *Structural Survey*, **1**(3), 254, 256–61.

ISE (1994) *Subsidence of Low Rise Buildings*; Institution of Structural Engineers (March).

# 8.4 Other soil shrinkage and foundation problems

**Learning objectives**

You should be able to distinguish between:

- subsidence due to wholesale, mining, clay and tree-induced movements in soil;
- other forms of foundation/soil failure mechanisms and symptoms.

In exceptional and/or prolonged dry periods the soil underneath the building may retain its moisture longer than that surrounding the building. The surrounding soil will shrink and lose its bearing capacity as it dries out; in prolonged drought this effect may extend to the depth of the foundations.

When this occurs the relative pressure of the damp soil underneath the building will exert an outwards pressure on the foundations and surround walls. Drying in the surrounding soil may also allow the foundations to settle as the bearing capacity of the soil deteriorates.

The effect on the building is twofold, and manifests itself as cracking. The walls move downwards and outwards and cracks may appear (usually passing through openings) as the walls are placed into tension. Inspection of the surrounding soil will probably indicate separation and cracking in the proximity of the building perimeter.

This problem produces similar faults to tree damage (which is usually focused on the side of the building nearest the tree).

## Other foundation problems

A range of localized problems in low-rise buildings can occur as a result of pockets of soil failure, direct failure in the foundations (including subsoil walls) or by a failure in combination of soil and foundations. The causes of such failures may be chemical attack due to chemical incompatibility between the soil and foundation (e.g. attack of concrete foundations by soil sulphates in solution), or even within the building fabric itself.

*The Technology of Building Defects.* Dr John Hinks and Dr Geoff Cook.
Published in 1997 by E & FN Spon, 2–6 Boundary Row, London SE1 6HN, UK. ISBN 0 419 19770 2

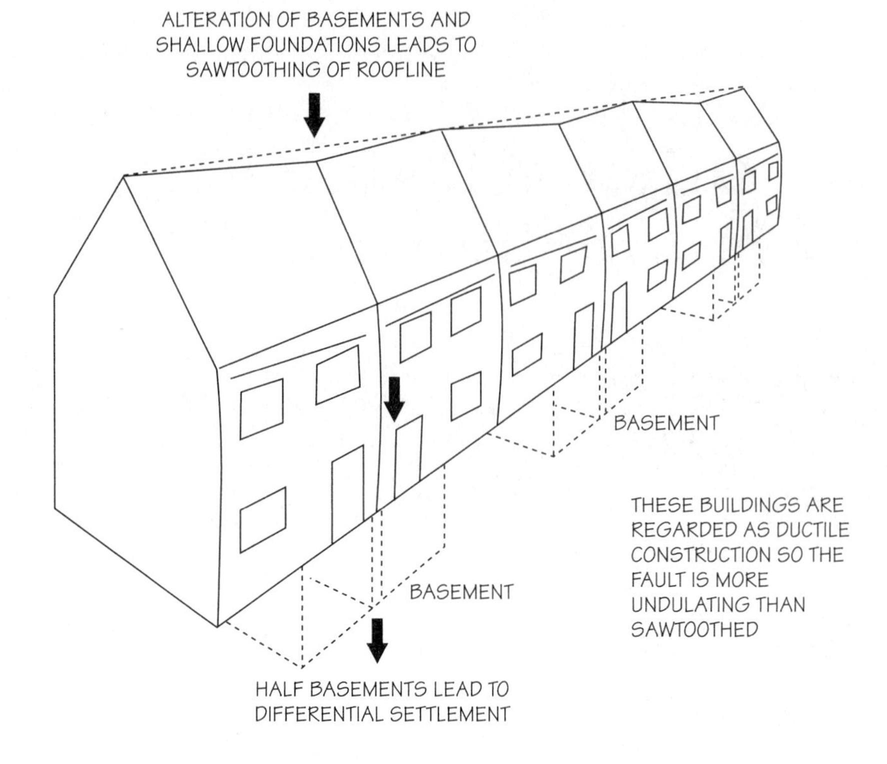


**Fig. 8.9.** Sawtooth undulations in terraced buildings.

**Fig. 8.10.** Severe gable end cracking. The crack width increases with height indicating a movement of the lower left-hand corner of the wall.

Alternative mechanisms include local erosion of soil, overloading of foundations or localized failure of soil strata/foundation support, and also redistribution of loads in the building structure.

The cracking/movement symptoms of foundation failure can differ from other causes of cracking such as clay soils in some generic ways, which may or may not appear in particular circumstances.

For instance, general foundation movement may be characterized by the presence of a relatively small number of cracks occurring at weak points in the structure. Whether such cracks cross the DPC, and are of uniform or tapering width can help distinguish the causes. Usually with foundation movement the cracks taper from top to bottom and exceed 2–3 mm width. In contrast to some material movement phenomena, the cracking will probably manifest itself inside and outside the building and produce distortion in elements which become out of plumb. Other indicators may include spreading of roof tiles or sticking of doors and windows (symptoms which can also occur with subsidence and differential movement caused by soil forces, note).

---

## ■ Discussion topic

- Produce a table to compare the causes and symptoms of cracking in buildings. Identify those factors and symptoms which allow different causes to be positively distinguished.

---

# Further reading

BRE (1991) *Why Do Buildings Crack?* Digest 361 (May), Building Research Establishment.

Shabka, G. and Kuhwald, K. (1995) Subsidence and the Associated Problems with Reference to Low-rise Housing. *Structural Survey* **13**, 28–35.

# 8.5 Basements and tanking

## Learning objectives

- You should be able to appreciate the causes of a range of defects associated with basements and cellars.
- In recognizing that a failure of the waterproofing function is a key defect, you should be aware of the influence of condensation and the deterioration of materials. Remember to consider the possible defects associated with any remedial works.

## General faults

Many of the general faults with buildings begin in cellars or roofs. Access to a cellar or basement may allow inspection of the ground floor including DPCs services, etc. Care is needed where the basement walls are damp and in contact with timber stairs since they may be liable to insect and fungal attack.

Older basements may be without any damp-proofing. Dry areas provided some protection by the construction of a retaining wall some distance in front of the basement wall. The open-area construction method allowed suspended timber floors to be constructed in the basements. These can decay or rot owing to inadequacies in the ventilating provision and/or leaking basements.

Remedial treatment to older basements which involved the rendering of the external face of the basement wall and replacing the brick floor with concrete can also fail owing to a lack of waterproofing integrity.

Loadbearing brickwork on concrete foundations was typical construction at the turn of the century. Where these were built under parts of urban Victorian terraces there is the potential for a sawtooth-shaped settlement of the terrace.

Ground movement due to clay heave does not usually extend to basement or cellar depth. Major subsidence due to mining or other activities may affect surrounding buildings.

Basements may be subject to flooding. This may be permanent or intermittent. Flooding directly into the basement can occur. Alternatively, water may pass through the basement structure. This groundwater may come

*The Technology of Building Defects.* Dr John Hinks and Dr Geoff Cook.
Published in 1997 by E & FN Spon, 2–6 Boundary Row, London SE1 6HN, UK. ISBN 0 419 19770 2

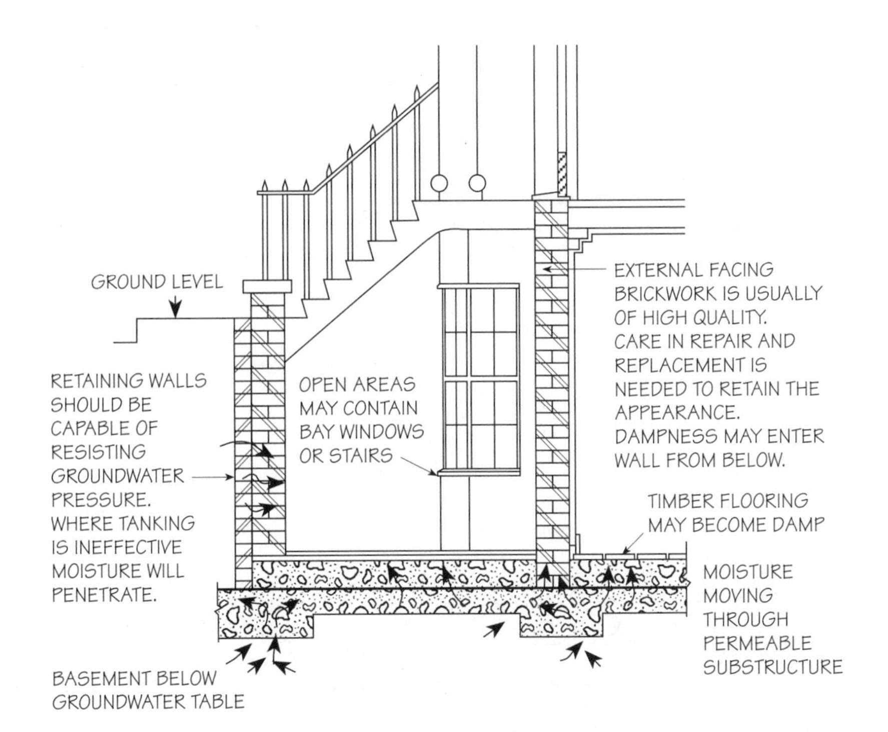


**Fig. 8.11.** Basement with open area – failure mechanisms.

from local watercourses which vary in depth with the seasons. Blocked or leaking drains and soakaways in the vicinity can cause increases in water pressure although fractured water mains can give rise to sudden and substantial increases.

## Common failures

Basements may fail because of a range of defects, including the following.

1. An incomplete water barrier due to failure of the tanking or an inadequate and incomplete link with the ground floor DPC. This may produce dampness around service entry points. Differential movement of the building may produce cracking. In addition, cracks may appear near structural, constructional or movement joints in walls and floors. The junction between floor and wall is particularly vulnerable.

   Patches of dampness high up in one place may mean that the tanking has failed at a point. The causes of this and more general water entry may not be straightforward to locate, particularly when the tanking is sandwiched within the basement wall.
2. The inability of the tanking to resist the hydrostatic pressure caused by water in the ground. Waterproof renders applied internally may be sufficiently impermeable but fail owing to a lack of bonding to the basement wall. Tanking within the basement wall can fail where the inner wall is unable to resist the hydrostatic pressure. This will increase with depth, unlike the applied load from soil. The ground drainage pipes commonly laid in the ground in front of the basement wall may become

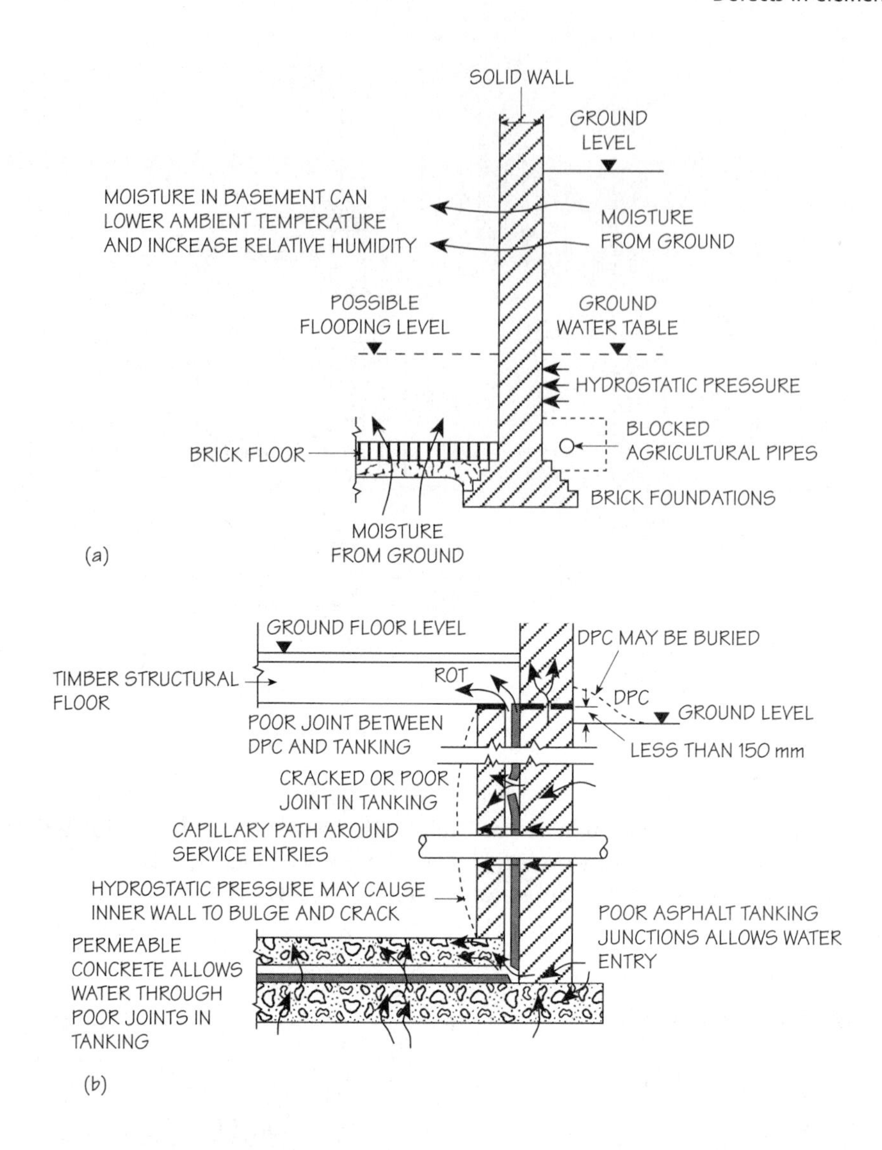


**Fig. 8.12.** (a) Older basement construction – failure mechanisms; (b) newer basement construction – failure mechanisms.

silted up, causing groundwater levels to rise. Early forms of tanking were applied to the external face of the basement wall.

3. Failure of tanking to bond with the background material.
4. Dampness due to condensation. This may occur on inner wall and floor surfaces, particularly in intermittently heated basements with poorly insulated walls and floors. Basements which are dry-lined may hide the surface condensation which forms on the inner surface of the structural wall. This may be due to failure or omission of the vapour barrier. Where the cavity between the dry-lining and the structural wall is not ventilated, or ventilated to the basement, there is an increased risk of condensation.

## Revision notes

- Inspect cellars or basements since many faults begin or can be seen from there. Ground settlement is unlikely to occur at basement depths.
- Older basements may be without any damp-proofing as open areas were commonly used. Remedial work to these could now be defective.
- Main defect is an incomplete water barrier causing dampness or flooding.
- Condensation can occur with intermittently heated basements. Can occur behind dry-lining used to cover remedial treatment, particularly when the cavity is vented with basement air.

---

## ■ Discussion topics

- Produce a list of features to examine when carrying out the inspection of a basement or cellar in a domestic building.
- Compare the failure mechanisms of basements having open areas with other types which have a waterproof rendering on the inside wall of the basement.
- The inside walls of a basement appear to be damp. The walls are constructed of brick with an asphalt tanking within a cavity in the wall. There are also random patches of severe dampness on the basement walls near the ceiling. Suggest an investigative procedure to examine these defects and discuss possible causes.
- Explain how the structure of modern basements is likely to produce different types of defect symptoms compared with older basements.

---

## Further reading

BRE (1973) *Repair of Flood Damaged Buildings*, Digest 152, Building Research Establishment, HMSO.

BRE (1993) *Damp-proofing Basements*, Good Building Guide 3, 2nd edn, Building Research Establishment, HMSO.

Cook, G.K. and Hinks, A.J. (1992) *Appraising Building Defects: Perspectives on Stability and Hygrothermal Performance*, Longman Scientific & Technical, London.

PSA (1989), *Defects in Buildings*, HMSO.

Richardson, B.A. (1991) *Defects and Deterioration in Buildings*, E. & F.N. Spon, London.

Staveley, H.S. and Glover, P. (1990), *Building Surveys*, 2nd edn, Butterworth-Heinemann, London.

CHAPTER 9

# Walls

# 9.1 Rising dampness and thermal pumping

## Learning objectives

- You should be able to understand the principles of rising dampness and identify the failure of a DPC or DPM as a major cause.
- You will be able to differentiate between the symptoms and causes of rising dampness and condensation in relation to walls and floors.
- You will be made aware of the concept of thermal pumping of the external envelope of a building.

## Rising damp – walls

Rising dampness should not occur, because DPCs have been compulsory since 1875, although occasionally they are omitted. The traditional DPC used courses of bricks and slates bedded in rigid cement mortar. These can be effective barriers to rising damp, although they may not resist water percolating downwards. Unfortunately movement failure may result in capillary-sized cracks. More brick courses may not give greater resistance to rising damp. Modern DPC material is flexible and is liable to a range of failure mechanisms. Joints are vulnerable regions and where thin, flexible DPCs are used the joints are commonly lapped. Where the length of lap is reduced, simple butt joints are used or gaps exist between strips of DPC; this can allow dampness to move upwards through the wall. The mortar joint may hide these defects.

The water drawn upwards into walls and floors in contact with the ground can contain dissolved salts. The height to which dampness can rise in a wall is dependent on a variety of factors including the pore structure of the material. Since bricks and concrete are generally porous, dampness can migrate through them. Where the inside of the building is heated, dampness can evaporate from interior surfaces leaving behind hygroscopic salts. These can absorb moisture from the air within the building, which can add to the general dampness problem. The assessment of moisture content by pushing the probes of an electrical resistance meter into a wall must take account of the effects of hygroscopic salts. Surface samples can be compared with samples from within the wall thickness. Where there is a high

*The Technology of Building Defects.* Dr John Hinks and Dr Geoff Cook.
Published in 1997 by E & FN Spon, 2–6 Boundary Row, London SE1 6HN, UK. ISBN 0 419 19770 2

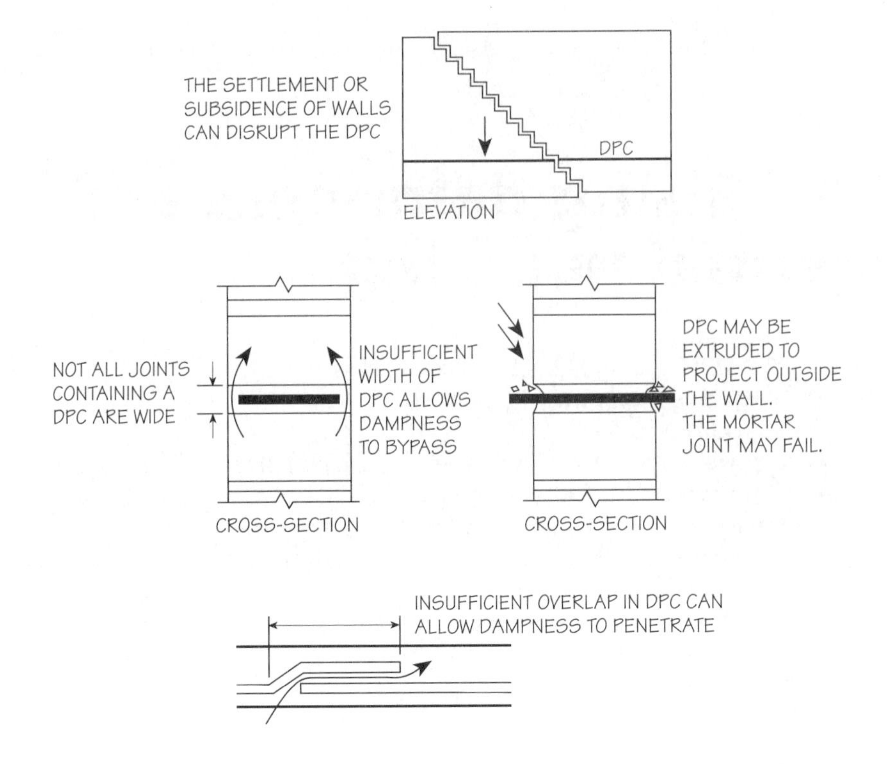


**Fig. 9.1.** General defects with DPCs.

moisture content in the surface sample, but little evidence of hygroscopic salts, then surface condensation may exist.

Rising dampness can be identified by a generally horizontal although irregular pattern of dampness. This is commonly associated with the discoloration of inner surfaces, which may rise to $>600$ mm above ground floor level.

If rising dampness occurs, the DPC material may have failed or have been inadequately jointed. Alternatively, the DPC may have been bridged. This would allow dampness to move around the DPC and move upwards through

**Fig. 9.2.** Symptoms of dampness within a building. The general shape and height of the staining suggests that this may be due to rising dampness. (K. Bright.)

the wall. There are various ways that the DPC can be bridged; they include the following.

## Failure to provide a continuous link between the DPM in the floor and the DPC in the wall

This may be due to a difference in level between a concrete floor slab and a brick course. Because DPCs and DPMs are commonly different materials, with different characteristics, they may fail at their junction. Where the DPC is above the DPM then dampness may migrate to a region behind the skirting board. Where the DPC is below the DPM dampness can migrate through the floor construction and into the wall. Where the DPM is formed by brushing on waterproof emulsions it may fail owing to inadequate thickness or general failures in workmanship. Polyethylene sheets of inadequate thickness, typically $< 0.25$ mm, can be punctured because of site conditions and allow dampness to enter. This can also occur where there is insufficient overlap and jointing of the polyethylene sheets.

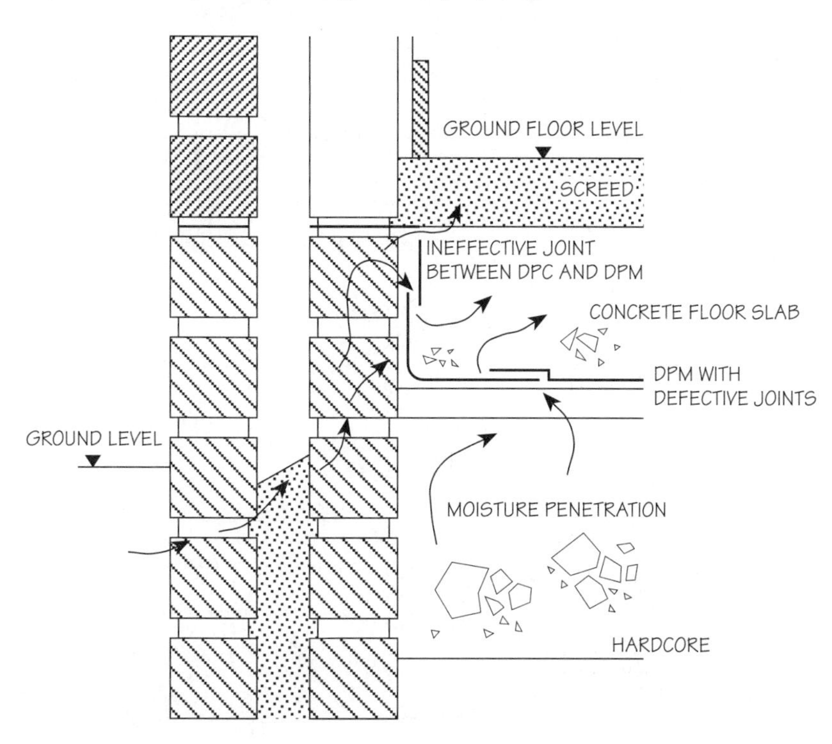


**Fig. 9.3.** Dampness penetration through ineffective joint between DPC and DPM.

## Mortar droppings at the base of a cavity wall

The cavity should extend at least 150 mm below the lowest DPC; where this has not occurred there is an increased risk of filling the cavity with mortar or other debris. Whilst there is the general assumption that mortar droppings are prevented from entering wall cavities by good workmanship or

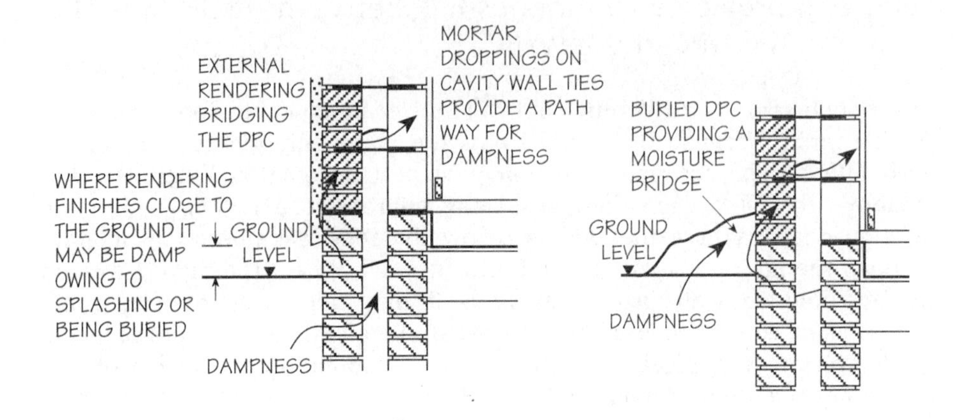


**Fig. 9.4.** Bridging the DPC.

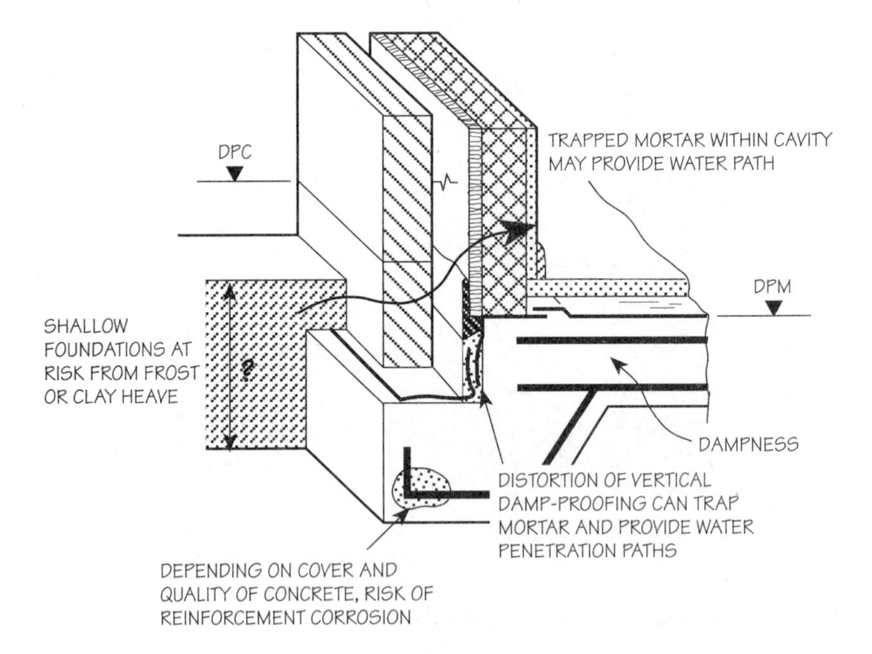


**Fig. 9.5.** Water penetration routes through raft foundation junction with walling.

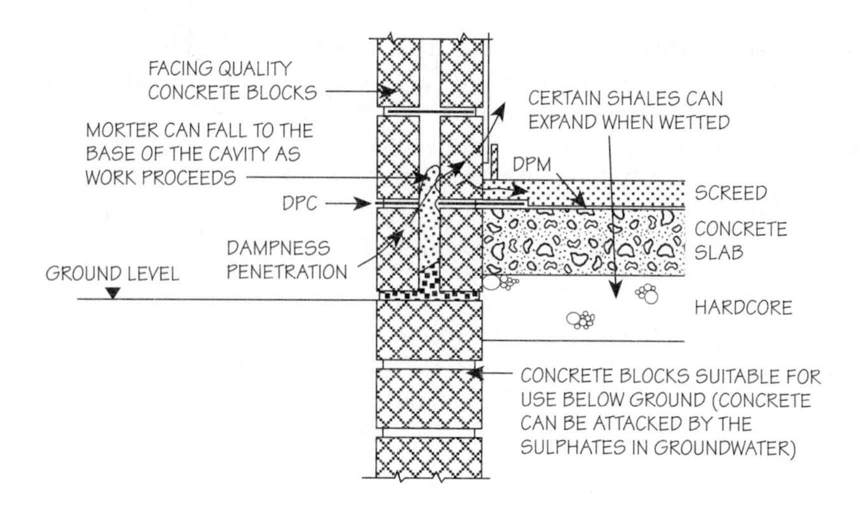


**Fig. 9.6.** Dampness penetration through mortar droppings at the base of the cavity.

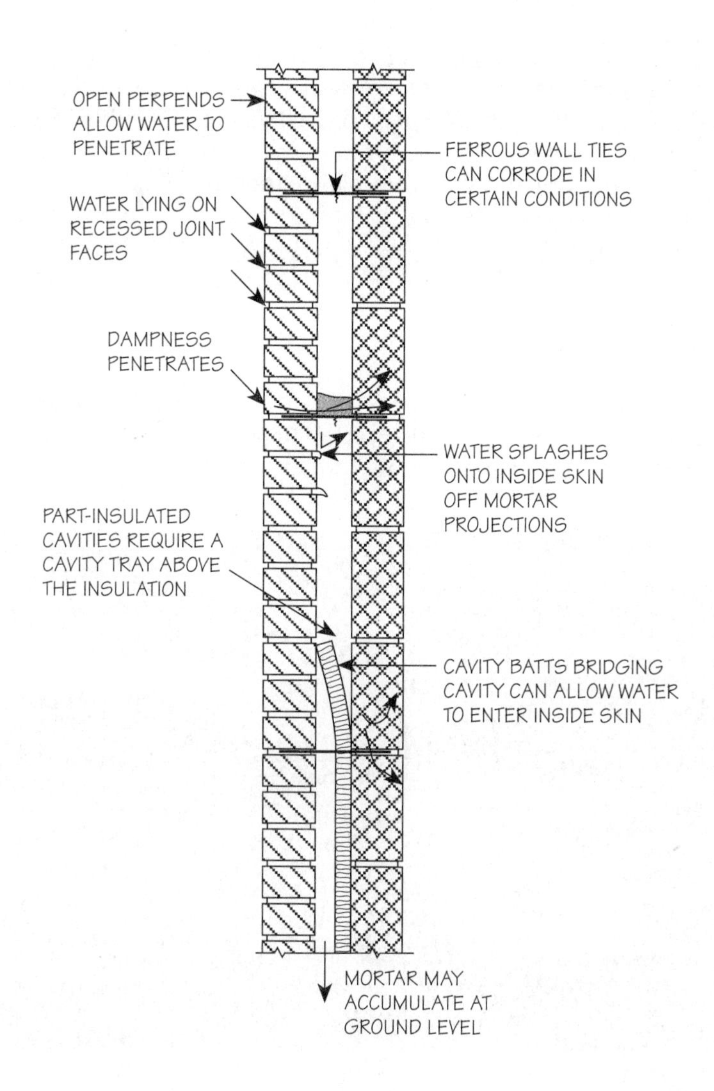


**Fig. 9.7.** Dampness penetration through cavities.

other measures, this may not occur in practice. Accumulations of mortar which extend above the DPC level can act as a conduit for rising dampness. Where DPCs are stepped between the outside and inside leaves of a cavity wall the cavity in this region is relatively small and therefore more easily filled with mortar droppings. A similar problem can occur with a concrete raft foundation.

The symptomatic dampness may appear on inner surfaces in a more irregular pattern than that associated with a wholesale omission or bridging of the DPC.

The practice of filling existing cavities with insulating material may reduce evaporation rates within the cavity and provide alternative pathways for dampness around mortar droppings. Remedial treatments, e.g. the injection of a chemical DPC, may also fail to stop a bridging route for rising dampness.

## Pointing or rendering over the DPC

The need for an acceptable appearance of the generally thicker mortar joint containing the DPC may have inadvertently provided a bridge for rising dampness. Slight movements and possible extrusion of flexible DPC material can cause the mortar joint to deteriorate. Severe extrusion of the DPC may be due to expansion of the concrete slab, since the flexible DPC material may act as a slip plane. This may have affected the integrity of the DPM under the concrete slab.

## Soil or external paving

The 150 mm minimum distance between ground and DPC level may disappear owing to gardening activities or pavings. Providing an effective DPC at door thresholds can be problematic.

**Fig. 9.8.** The partially buried air brick suggests that the DPC is also close to ground level. (K. Bright.)

Soil can carry moisture to the outside face of the wall and retain run-off water close to the wall or in contact with it. Vegetation may encourage the growth of external moulds. Whilst this may allow dampness to rise into the external leaf of cavity walls, it is potentially more serious for solid walls. Paving at or above DPC level can allow rain splash to wet further areas of the wall. Weepholes at low level may provide a route for water to enter the cavity, either by splashing or their being buried.

The wall immediately below the DPC may be damp for long periods. Frost attack can occur and there is an increased risk of sulphate attack.

### Vertical DPCs

Ineffective DPCs may be related to the position of the door or window frame in the wall. Windows or doors which are deeply recessed may place the window-to-wall joint behind the vertical DPC. This can lead to direct water entry, particularly where DPCs are poorly positioned in the structure. The surfaces of the DPC can become a capillary route for water penetration. Where windows and doors are towards the external face of the wall they are more vulnerable to weathering.

## Rising dampness – floors

Solid concrete floors, where the DPM has failed or was omitted, can deteriorate owing to sulphate attack from the hardcore. This can cause cracking, which can also occur when the hardcore material swells. Unhydrated lime and magnesia in slag from steelworks, refractory bricks and types of shale can expand when damp. The deflection and cracking of the concrete may also fracture the DPM, where this is laid over the concrete. Poorly compacted hardcore may settle, causing the concrete slab to become unsupported. Shear and bending cracking can then occur, which may also rupture the DPM. Localized damp patches on concrete floors may also be due to leaking buried pipework.

Failure mechanisms of the DPC on sleeper walls in suspended timber floors are similar to those in cavity walls. Where air bricks are blocked or buried the required natural ventilation of the floor may not occur; this can lead to high moisture contents in the timber and decay. Owing to the differences in level between air bricks and the DPC it may be possible to bury the air brick and still retain a minimum 150 mm clearance between ground level and the DPC. Alternatively the air bricks and the DPC may be buried.

## Condensation

Human activity produces moisture; it is suggested that this may be $>14$ litres per person per day. The majority of this moisture is produced in the kitchen and bathroom. Although the chronic problems of condensation associated with paraffin heaters in the 1960s are rare, condensation continues to cause concern. The need to conserve fuel, increase insulation and reduce natural ventilation caused by gaps in and through the building structure has increased the risk of condensation. The trend to increase the amount of insulation in walls has moved the areas of lower temperature towards the outside. This suggests that modern external walls are likely to have their external surfaces exposed to lower temperatures for longer periods. This may influence the long-term durability of the wall.

In addition, the general domestic lifestyle has demanded intermittent heating of the building. The corresponding changes in temperature and moisture content of the air can increase the incidence of condensation. Warm, humid air inside a building may have a greater vapour pressure than external air and therefore may migrate through the structure. When this air cools to below its dew point temperature, condensation can occur. This can occur on relatively cool surfaces or within vapour-permeable materials to produce surface or interstitial condensation respectively. Effective vapour barriers can reduce the incidence of interstitial condensation. Whilst it is possible to predict the occurrence of interstitial condensation when the nature of the structure and the design conditions are known, this commonly assumes steady-state conditions. Where heating is intermittent several other factors involving the thermal inertia of the fabric need to be considered. Interstitial condensation within fibrous insulating materials will substantially reduce the thermal resistivity of the material and significant amounts of moisture may be present.

Increasing the moisture content of porous materials is likely to increase their thermal conductivity. The resulting increase in heat flow from within the building can increase the incidence of interstitial condensation and use more fuel to maintain the same internal comfort conditions.

Localized areas of high thermal conductivity form 'cold bridges'. These areas of high heat flow and relatively lower temperatures can become sites

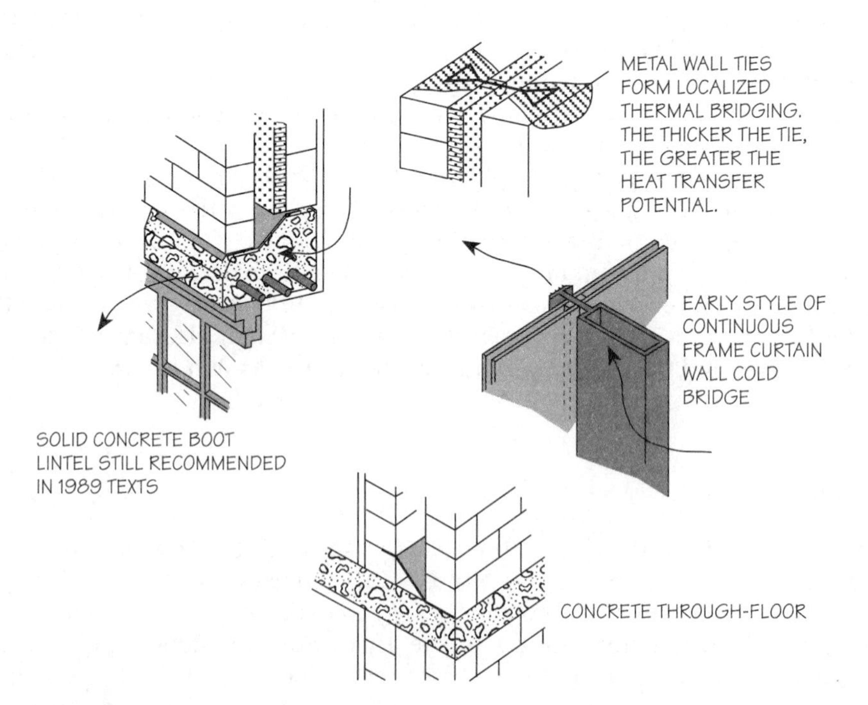


**Fig. 9.9.** Simple forms of cold bridge.

for condensation. This may be due to a variety of factors, including mortar droppings on wall ties in cavity walls, concrete lintels and columns. The areas around the edges of concrete ground- and upper-floor slabs can also act as cold bridges. The presence of moisture will be intermittent, compared to a loss of integrity between the DPM and DPC.

Where there is condensation there may also be ideal conditions for mould and fungus growth. In general, relative humidity levels of $> 70\%$ are required for fungal growth, with levels of $> 80\%$ for sustaining growth. These relative

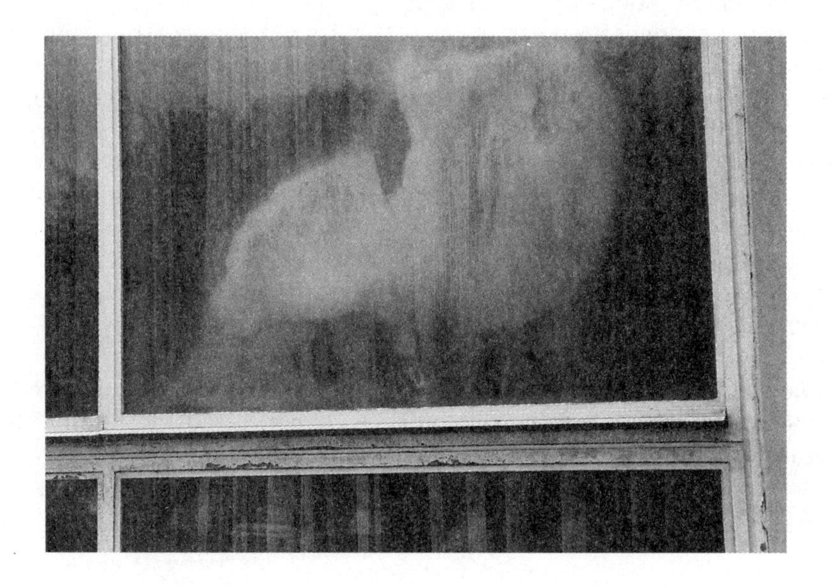

**Fig. 9.10.** Incidence of condensation within a double-glazed window.

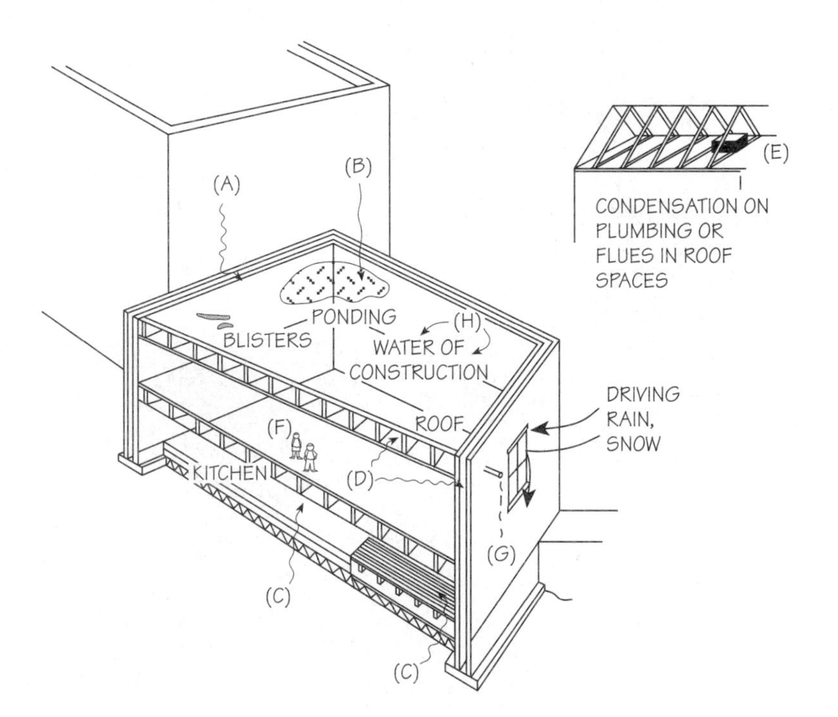


**Fig. 9.11.** Moisture in buildings: sources and problems. Extrinsic sources and routes: (A) permeating dampness from adjacent buildings; (B) leakage of roofs; (C) soils.
Internal sources and routes: (D) interstitial condensation; (E) condensation within spaces; (F) people.
Intrinsic sources and routes: (G) plumbing overflow – runback; (H) entrapped moisture.

humidity levels can also be provided by penetrating dampness, rising dampness or the water which is part of the construction process. This may be stored within the wall or have evaporated from other walls to add to the general relative humidity level of the air in the building.

Fungi tend to be dark-coloured and unattractive. They can also cause the deterioration of furnishings and finishes. Wet timber can rot.

Windows can be a particular risk; some used condensation channels and tubes to remove the condensation, although this placed the wall surface below the cill at risk from frost attack.

Interstitial condensation can occur within timber window frames where a vapour barrier coating is not provided. This can lead to wet rot. Where the seals to double-glazed units have failed then condensation can occur in the glass cavity.

Mechanical ventilation of kitchens and bathrooms may transfer the incidence of mould growth from interior surfaces to surfaces of the fan and any associated ductwork.

## Thermal pumping

This is a term devised by Dr W. Allen and is caused by the changes of the external temperature and the difference between it and the internal temperature within the building. During cold or wet weather, the roof or walls of the building become cold. Consequently there can be a lower vapour

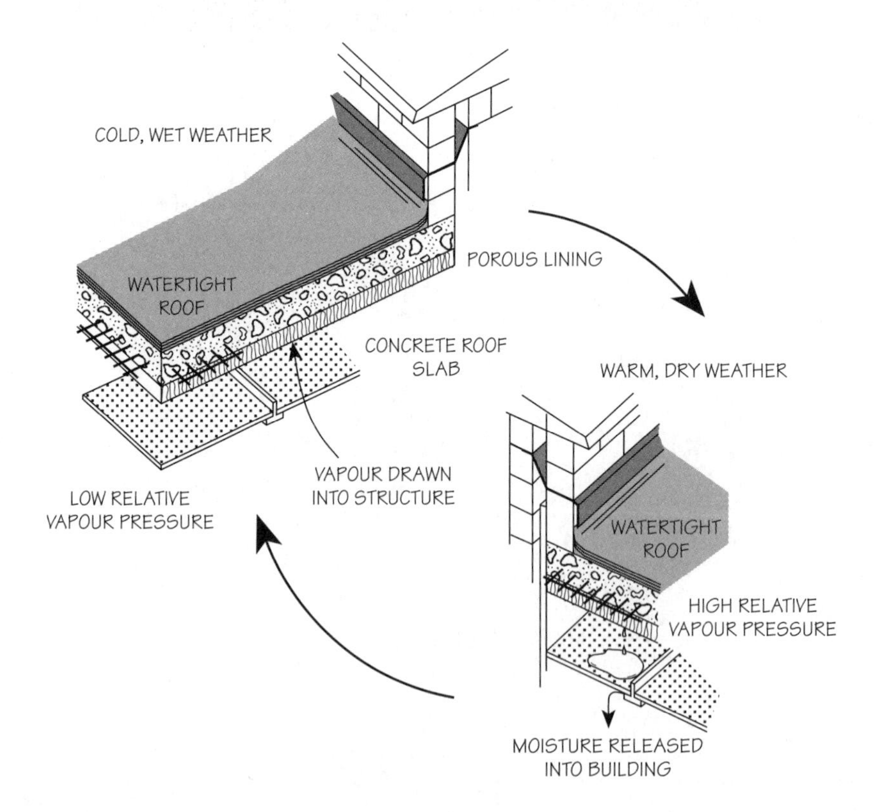


**Fig. 9.12.** Thermal pumping mechanism. (W.A. Allen.)

pressure in the external envelope thickness than inside the building. Moisture can be drawn into the layers of the external envelope from the inside of the building, and held there. Humid atmospheres are a particular risk.

During warm weather the roof slab or external envelope can substantially increase in temperature, particularly so where it is insulated at the rear. This increases the vapour pressure of the trapped moisture, and where it exceeds the internal vapour pressure, moisture is forced back out into the interior of the building. Moisture then appears, perhaps as drips from a ceiling, at times when conventional leakage is unlikely.

## Revision notes

- Traditional, brittle brick or slate DPCs in walls can crack to allow dampness to rise through capillary action. Modern, flexible DPCs can be extruded owing to load. Joints can fail owing to inadequate laps or sealing. DPCs can be bridged in a variety of ways. An ineffective DPM to a solid concrete floor allows sulphates to attack the concrete. Suspended timber ground floors may have the DPC and/or air bricks buried.
- Rising dampness in a wall can be identified by a generally horizontal although irregular pattern of dampness. This can contain hygroscopic salts, therefore care is needed when using an electrical resistance probe meter. Where a high moisture content exists on the wall surface, but little evidence of hygroscopic salts, then surface condensation may exist. Condensation is more likely when humidity is high, heating is intermittent, vapour barriers are omitted, cold bridges occur and the external envelope is cold.
- Thermal pumping is where moisture which is drawn into the outer structure when it is cold outside is drawn back into the building when the outer structure warms up.

## Discussion topics

- Compare the long-term effectiveness of rigid and flexible DPCs in masonry walls.
- Describe three failure mechanisms which can result in rising dampness in a masonry cavity wall.
- Discuss the factors which influence the incidence of condensation in buildings and the implications for the long-term durability of internal materials and finishes.
- Explain the influence of intermittent heating on the incidence of interstitial condensation on an external wall.
- Describe an inspection procedure for a wall in order to differentiate between dampness caused by condensation and that caused by failure of the DPC.

- Discuss the durability implications of highly insulating external walls.
- Explain the concept of 'thermal pumping' in relation to the external wall of a timber-framed building.

---

## Further reading

BRE (1965) *Damp-proofing Solid Floors*, Digest 54, Building Research Establishment, HMSO.

BRE (1981) *Rising Damp in Walls: Diagnosis and Treatment*, Digest 245, Building Research Establishment, HMSO.

BRE (1983) *Walls and Ceilings: Remedying Recurrent Mould Growth (Design)*, Defect Action Sheet 16, Building Research Establishment, HMSO.

BRE (1983) *Substructure: DPc's and DPM's – Installation (Site)*, Defect Action Sheet 36, Building Research Establishment, HMSO.

BRE (1989) *Suspended Timber Ground Floors: Remedying Dampness due to Inadequate Ventilation (Design)*, Defect Action Sheet 137, Building Research Establishment, HMSO.

BRE (1991) *Tackling Condensation*, BRE Report 174, Building Research Establishment, HMSO.

BRE (1992) *Interstitial Condensation and Fabric Degradation*, Digest 369, Building Research Establishment, HMSO.

BRE (1993) *Damp-proof Courses*, Digest 380, Building Research Establishment, HMSO.

Cook, G.K. and Hinks, A.J. (1992) *Appraising Building Defects: Perspectives on Stability and Hygrothermal Performance*, Longman Scientific & Technical, London.

PSA (1989) *Defects in Buildings*, HMSO.

Ransom, W.H. (1981) *Building Failures: Diagnosis and Avoidance*, 2nd edn, E. & F.N. Spon, London.

Richardson, B.A. (1991) *Defects and Deterioration in Buildings*, E. & F.N. Spon, London.

Staveley, H.S. and Glover, P. (1990) *Building Surveys*, 2nd edn, Butterworth-Heinemann, London.

# 9.2 Stone walls

**Learning objectives**

You should understand the mechanisms and symptoms of defects in stone walls, including:

- chemical and physical attack;
- fixing and movement problems.

Stone was one of the earliest forms of natural wall material with acceptable performance characteristics. A variety of constructional methods performed well, since the material was well understood.

The principal mechanisms of deterioration in stone walls are material degradation (including pointing) and other structural failures such as movement cracking. The causes are the design/construction detailing and use, and/or the chemical/physical degradation of the material. Critical defect areas include weathering of the stone, corrosion of ties and cramps and pattern staining.

Since most traditional stone buildings involve heavy, large stones lying in relatively weak beddings, the buildings tend to be ductile and absorb the

**Fig. 9.13.** Delamination of stone surface. The initial, developing and advanced stages are clearly shown. (K. Bright.)

*The Technology of Building Defects.* Dr John Hinks and Dr Geoff Cook.
Published in 1997 by E & FN Spon, 2–6 Boundary Row, London SE1 6HN, UK. ISBN 0 419 19770 2

relatively slow, moderate movements well (this can be a problem with arched construction!) Filled stone walls may suffer internal instabilities, however, which are particular to this type of construction.

## Crystallization

The most widespread problem arises from the crystallization of soluble salts (the mechanisms of which are dealt with in the 'Materials' part of this book). This can cause sufficient pressures to build up behind the face of the stonework to fragment the stone, leading to localized splitting or spalling at its surface.

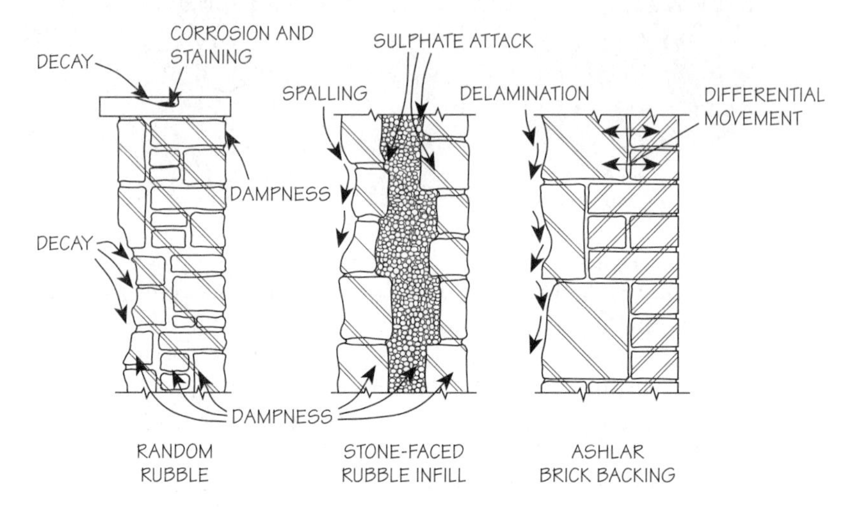


**Fig. 9.14.** Defects in stone walls.

## Erosion

It is common for this mechanism to cause gradual erosion, such as in limestones, aided by the washing of rainwater. The sedimentary deposition of stones can exacerbate such problems, as will details which expose the stone to run-off. The rainwater run-off on stone buildings may produce staining as it leads to the transportation of salts to other locations, leading in turn to secondary attack on other materials. Drying out of the stonework can itself be destructive, more so if a dense and relatively impermeable mortar leads to most of the evaporation (and therefore moisture-transport-related erosion) occurring from the stone surface.

## Fixing failures

In addition to direct damage to the stone, any ferrous fixings which corrode will cause disruption upon swelling. The issue of fixing security has been raised also by some failures in modern thin stone claddings, some of which have inhibited a relatively high thermal and moisture movement behaviour. In extreme cases, the attachment of the stone, which itself is a source of problems where the stone is attached too rigidly and without capacity for movement, has produced localized cracking and spalling at connections (or

**Fig. 9.15.** General weathering of stone wall. The pattern of attack has been influenced by the position of replacement stone. (K. Bright.)

**Fig. 9.16.** The staining of this external façade is more pronounced in the internal corner, where the effect of rain washing is less marked.

even dislocation of the stone panels). Hence thin stone claddings should be checked for adequacy of stress-relieving movement joints.

## Chemical attack

Specific chemical attack can be disruptive. Carbon dioxide in the air, which supports carbonation in concrete, can lead to reactions involving the calcium carbonate which binds calcareous sandstones. In turn this affects the stability of the stone and causes erosion.

## Frost attack

Porous stones may suffer frost attack (see 'Materials') and surface spalling as chemical depositions block the pores and cause localized spalls, map cracking or detachment of the surface due to moisture/frost back pressure.

## Organic attack

There may also be a build-up of organic growths on the surface of the stone. Some lichens exude acids which directly attack the stonework, in addition to problems with exposure to acid rainwater (this is most acute in urban areas).

## Movement

The ductility of stone buildings can frequently accommodate considerable distortion through dimensional readjustment. In cases of significant differential movement there may be problems as the capacity of the stone masonry to accommodate movement is overstretched. In such instances cracking of the joints and/or the stone itself can occur. Cracking may also occur as a result of poor attention to detailing and/or construction. The causes of such defects include poor bedding, which creates localized rigidities causing the stone to become overstressed, also background instability which may arise with poor core fill to walls. Such fills were usually poor-grade random rubble with a greater number of less precise joints than the facing masonry. The distribution of stresses by movement in the fill is therefore possible, causing a weakening to the facing by reducing the background support/rigidity. Some rubble masonry was only filled with loose stones and is therefore less able to cope with defects in the surface masonry. The most basic of rubble walls usually have little capacity for overloading.

Where timbers are built into the walls, including bressummers, deterioration of the timber can become quite advanced, leading to the potential for localized collapse.

### Revision notes

- Several interactive causes of defects in traditional stone walls.
- Crystallization of soluble salts in the pore structure, leading to spalling.
- Frost attack due to moisture ingress/entrapment, producing localized spalling, splitting or map cracking.

- Erosion of stonework by abrasion, water run-off and/or chemical attack.
- Chemical attack, which may be due to ingress of environmental pollution (including salts) or incompatibility between stones, leading to spalling or erosion.
- Organic attack which may be a secondary defect or a primary cause of failure (e.g. acidic lichens).
- Fixings, which may fix the stone too rigidly (especially problematic in association with large panels and under-design of movement joints; this problem extends to bedding of stone in hard, dense mortars).
- Internal instability of core fill in ashlar, or particularly rubble, walls.
- Movements in the structure, especially differential movement.

---

## ■ Discussion topics

- Identify the role of detailing in the deterioration of stone and make a checklist for the inspection of stonework.
- Why should limestones not be placed above sandstones?
- Compare and contrast the characteristic defects in stone walls with other masonry walls.

---

## Further reading

Ashurst, J. and Ashurst, N. (1994) *Stone Masonry* Practical Building Conservation Series, English Heritage Technical Books.

BRE (1990) *Decay and Conservation of Stone Masonry,* Digest 177, Building Research Establishment.

CIRIA (1994) Structural renovation of traditional buildings. *Construction Industry Research and Information Association Report 111.*

Stone Federation (1995) *Movement Joints in Natural Stonework Data Sheet.*

# 9.3 Concrete panel walls

## Learning objectives

You should understand:

- the various mechanisms involved in producing characteristic sets of defects which can affect many types of concrete panel walling;
- the role the constructional detailing plays in achieving good performance or deficiency in use.

## Water penetration

Precast concrete panels may behave in a similar manner to solid walls with respect to water penetration. Surface rendering or other treatments will alter the behaviour of the panel with respect to water absorption. Some textured and otherwise untreated surfaces will have a relatively high propensity to absorb rainwater, and in such instances the quality of reinforcement cover will be critical to the durability of the panel as a whole.

## Water run-off and sealants

It is worth remembering that although the absorbency of concrete panels can reduce the load on joints of rainwater run-off it does not eliminate it; and in the cases of high-rise buildings the exposure to driving rain can be significant. The performance of sealants is therefore of critical importance to the overall performance of the walls (see elsewhere in this book for general sealant defects).

## Fixings

In order for the connection of concrete cladding panels to be successful it is necessary that the structure, the panel and its fixings have been considered together. Problems can arise with each of these factors, but also with incompatibilities between them. Some early examples of precast concrete

*The Technology of Building Defects.* Dr John Hinks and Dr Geoff Cook.
Published in 1997 by E & FN Spon, 2–6 Boundary Row, London SE1 6HN, UK. ISBN 0 419 19770 2

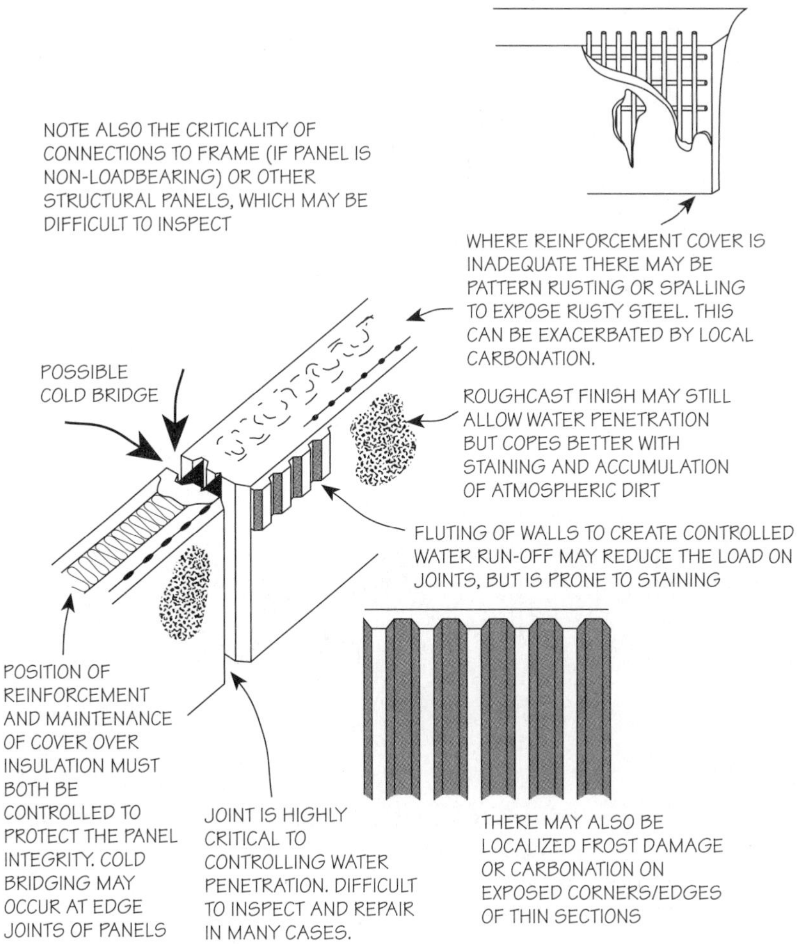

**Fig. 9.17.** Schematic view of characteristic defects in solid concrete wall panels.

cladding failed because the details were unbuildable in the context of site conditions and standards. There were also examples of profound misunderstanding about the constructional methods and criticality of workmanship which compounded these issues and led to a spate of problems associated with fixings. Examples of poor or omitted fixings have been found which can leave the panel and/or building stability in question.

Fixings will either tie the structure to the building or other adjacent structural panels, or restrain the panel from wind movement. Limitations in adjustment and movement size will produce alignment problems which may in turn cause problems with the accommodation of thermal and/or moisture movement, and also resistance to water penetration.

There are distinctions between the function and operation of top-hinge and base-support fixings which should be borne in mind when assessing the deficiency or otherwise of fixing. However, the appearance of movement symptoms due to localized over-attachment coupled with movement stresses is likely to appear at the points of least fixity. Hence base-supported panels are likely to exhibit movement at their head. Heavily fixed units will exhibit their movement at their joints (imposing loads on the sealants and

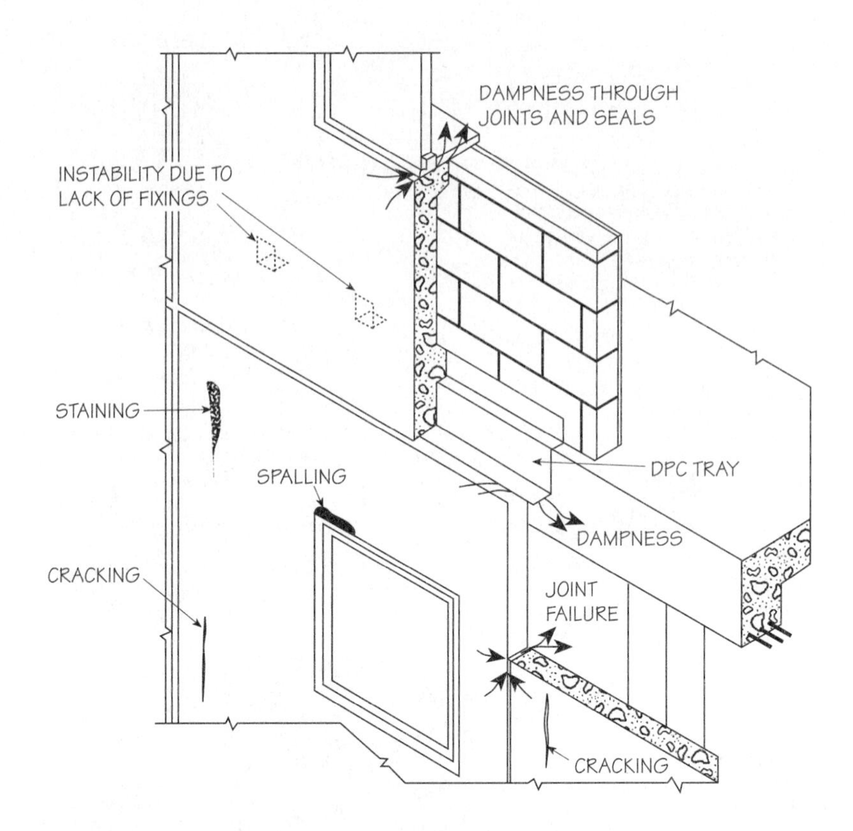


**Fig. 9.18.** Schematic view of characteristic defects in pre-cast concrete walls.

waterproofing). Note that either type of fixing will deteriorate faster if overstressed owing to panel inaccuracy during assembly.

Aside from the operation of sound fixings, there are also potential problems arising from the rusting of fixings, directly in terms of residual loadbearing capacity and resistance to induced stresses, and indirectly in terms of inducing expansive stresses in the panels and/or frame. The use of combinations of dissimilar metals can give rise to electrolytic corrosion. Faults with fixings may give symptoms on the cladding, such as pattern staining or cracking at or around attachment points.

In addition to the problems with fixings, there can also be defects arising from the bearings upon which panels rest. These may occur in terms of the accuracy of fit of panels and any adjustment made to accommodate this vertical inaccuracy during construction (such as hacking off the floor surface or addition of packing material). Horizontal inaccuracy can mean that in some cases the bearing may be reduced by cumulative inaccuracy to very small dimensions – leaving the support of the panel as inadequate and vulnerable to any movement in the structure. In fires the hogging which occurs as a result of the temperature build-up can cause a reduction in bearing greater than some reduced residual bearing sizes.

Hence one of the key issues in appraising defects with claddings is assessing the balance between residual security of connection and all the various causes of potential dislocation or distortion. In some cases there has been sufficient evidence amassed to designate constructional forms in connection with their generic and latent defects. In such cases the inspection of the building for otherwise unseen defects may need to extend to the

**Fig. 9.19.** The exposure of poor or non-existent continuity of reinforcement between precast concrete column sections.

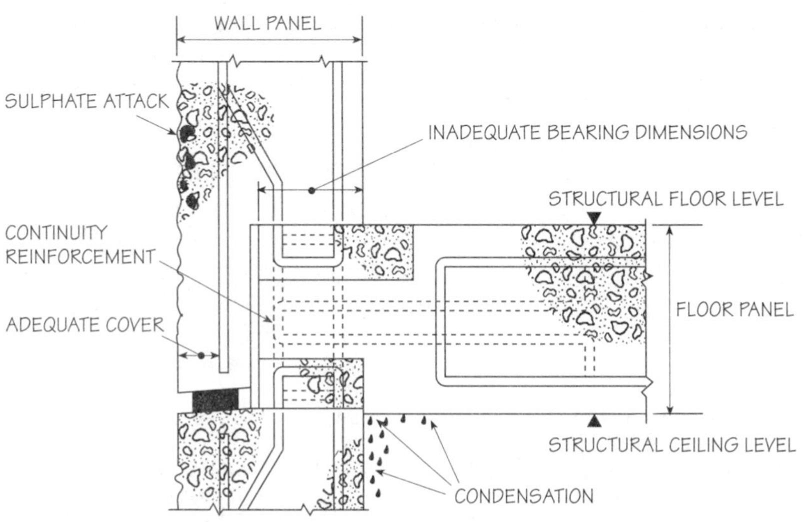


**Fig. 9.20.** Precast concrete walls – problems at external wall-to-floor junction.

consideration of such defects. In addition, the generic defects described above will have a greater incidence and severity in particular locations in the building. Such locations include parapets, corners or joints between different elements/materials.

## Revision notes

- Solid panels may behave similarly to traditional solid walls with respect to water penetration.
- Criticality of connections to frames or other structural panels.
- Problems with accuracy of panel dimensions and underprovision for movement tolerance can mean transfer or accumulation of stresses under high-movement conditions.
- Risk of surface rust staining or spalling if steel reinforcement corrodes. This is potentially high risk if reinforcement cover was inadequate. Positioning of insulation can exacerbate this.
- Joints are critical to prevent water ingress, but vulnerable to sealant deterioration and damage if insufficient movement accommodation was left between panels.
- Cold bridging may occur at perimeter of insulation and around joints.
- Early forms may suffer from progressive collapse risk, depending on design and subsequent stiffening and typing back.
- Possibility of surface staining and water run-off staining.
- Risk of carbonation (which also increases rusting risk in reinforcement).

## Discussion topics

- Produce an illustrated discussion of the role of fixings in failure of concrete panel walls. What are the likely symptoms?
- Research the issue of designated defects in concrete panel housing. What symptoms were designated and why? Produce a table of symptoms and defects, and review/analyse the scale/seriousness of the problem.

## Further reading

ACA *Fixings for Precast Concrete Panels*, Data Sheet 1, Architectural Cladding Association.

BRE (1974) *Wall Cladding Defects and Their Diagnosis*, Digest 161, Building Research Establishment.

BRE (1979) *Wall Cladding: Designing to Minimise Defects Due to Inaccuracies and Movements*, Digest 223, Building Research Establishment.

BRE (1980) *Fixings for Non-Loadbearing Precast Concrete Cladding Panels*, Digest 235, Building Research Establishment.

BRE (1982) *Reinforced Concrete Framed Flats: Repair of Disrupted Brick Cladding*, Defect Action Sheet DAS 2/82, Building Research Establishment.

# 9.4 Problems with timber-framed construction

## Learning objectives

- You should be able to understand the origins of a range of defects associated with timber-framed buildings. The text will develop this understanding by an explanation of the symptoms of certain defects.
- Moisture movement into and through a timber-framed building will be examined to enable you to describe the processes which increase the risk of deficiency.

## Timber-framed buildings – general problems

Timber-framed construction, whilst having the advantages of speed of construction and good thermal insulation, can be subject to a range of defects. These include the defects associated with the use of structural timber and, because of the obvious fire risk of this form of construction, a failure to provide fire barriers or stops. Controlling sound transmission can be problematic. Since production of timber-framed buildings can be industrialized, defects associated with tolerance and fit can also occur.

## Timber problems

Where there is inadequate protection of the timber frame during construction there is a risk of a high moisture content being present in the timber. The subsequent drying out of the timber during occupancy can cause distortion of the frame. Exposed metal fixings, including nails, may corrode. Where chipboard is exposed to moisture it may expand irreversibly, causing significant loss of strength.

Since the moisture movement of timber across the grain can be significant, timber panels are likely to expand and contract in the vertical direction. The initial drying out of the panels may produce a 5 mm reduction in height and where this and a possible 3 mm expansion of the clay brickwork used as an external cladding cannot be accommodated, distortion and cracking may occur. Although the amount of movement may be

*The Technology of Building Defects.* Dr John Hinks and Dr Geoff Cook.
Published in 1997 by E & FN Spon, 2–6 Boundary Row, London SE1 6HN, UK. ISBN 0 419 19770 2

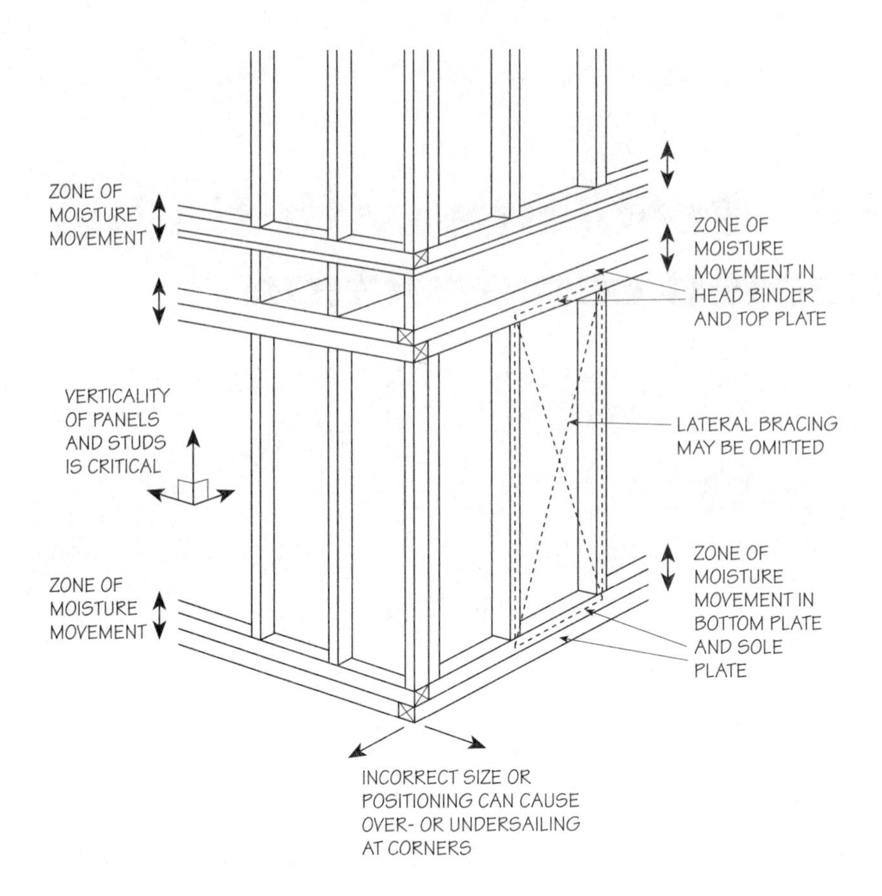


**Fig. 9.21.** Timber frame – movement problems.

substantially less, a failure to accommodate movement around window or door openings can distort frames.

Where inappropriate timber is used, this may be more likely to be affected by fungal decay. Inadequate preservative treatment can also reduce the durability of the timber frame and any associated plywood sheathing. Certain types of preservatives can be almost colourless and odourless, complicating identification. Chemical analysis may be required. Panels or associated structural timber which is modified on site can be vulnerable to decay, particularly where new end grain is exposed and untreated.

Obvious symptoms of rot in a timber frame usually appear when it has reached an advanced stage. There may be evidence of movement in the frame or between the frame and its attached components. This may be due to rot or structural failure.

There were also potential problems with the mismatch of timber standards of the country of origin and British Standards. There are many different timber standards and some may provide acceptable timber for construction only in the country of origin. This may be due to climate or construction technique.

The cutting away of parts of the timber frame in order to overcome inaccuracies in positioning or accommodate pipes or cables may weaken the structural integrity of the frame. This can be a particular problem with stressed-skin construction. Relatively large access spaces for services may weaken stressed-skin floors. Cutting through stressed-skin external panels may reduce the lateral restraint of the frame.

## Panel assembly

Inaccurate setting out of the commonly concrete base may lead to differences in level and size. Inadequate packing may cause distortion of the frame and where the frame is larger than the base then the cavity width may be reduced. This can lead to rain penetration. Alternatively the cavity width may be increased, adversely affecting the bond provided by the cavity wall ties.

The sole plates of timber panels may be inadequately fixed to the base. Where provided, the fixing may penetrate the DPC and allow dampness to occur under the sole plate. Poorly fixed sole plates may have structural implications since the imposed loads from the building are transferred to the foundations through the sole plates.

The factory-made framed panels are commonly nailed together, with the lateral restraint provided by plywood or chipboard panels glued or nailed to the frame. Adhesives can suffer from a range of defects, e.g. adhesive may have been omitted or incorrectly cured. Even when they have been applied under factory conditions they may still be inadequate.

The fixing between panels may be defective. The nailing schedules may have been ignored or nails simply omitted. The binder rails, which may be essential for the structural integrity of the frame since they fix adjacent panels together, may be inadequately nailed or have insufficient overlap to provide structural continuity.

## Moisture and thermal problems

The DPC may be omitted under sole plates. Where concrete slabs are not screeded there may be discontinuity between the DPM under the slab and the DPC under the sole plate. The underside of the sole plate may become damp, which increases the risk of fungal decay.

Inadequacies in the DPC provision around external openings may allow rain penetration. The nature of these inadequacies is influenced by the type of external cladding. An external skin of brickwork which forms a cavity construction can allow rain penetration in similar ways to that of a masonry cavity wall. Cavity ties which are not nailed directly to studs or plates may be structurally inadequate. 'L'-shaped ties may not provide structural integrity where the nailing at the end of the vertical leg of the tie allows the leg to distort, rather than resist the separation of the brickwork and timber panel.

Terraced or semi-detached houses may have different floor levels, leaving the lower section of a party wall exposed to penetrating damp from the ground where vertical damp-proofing is omitted or inadequate.

Cavity fire barriers bridge the cavity and may require a vertical DPC between the barrier and the external cladding.

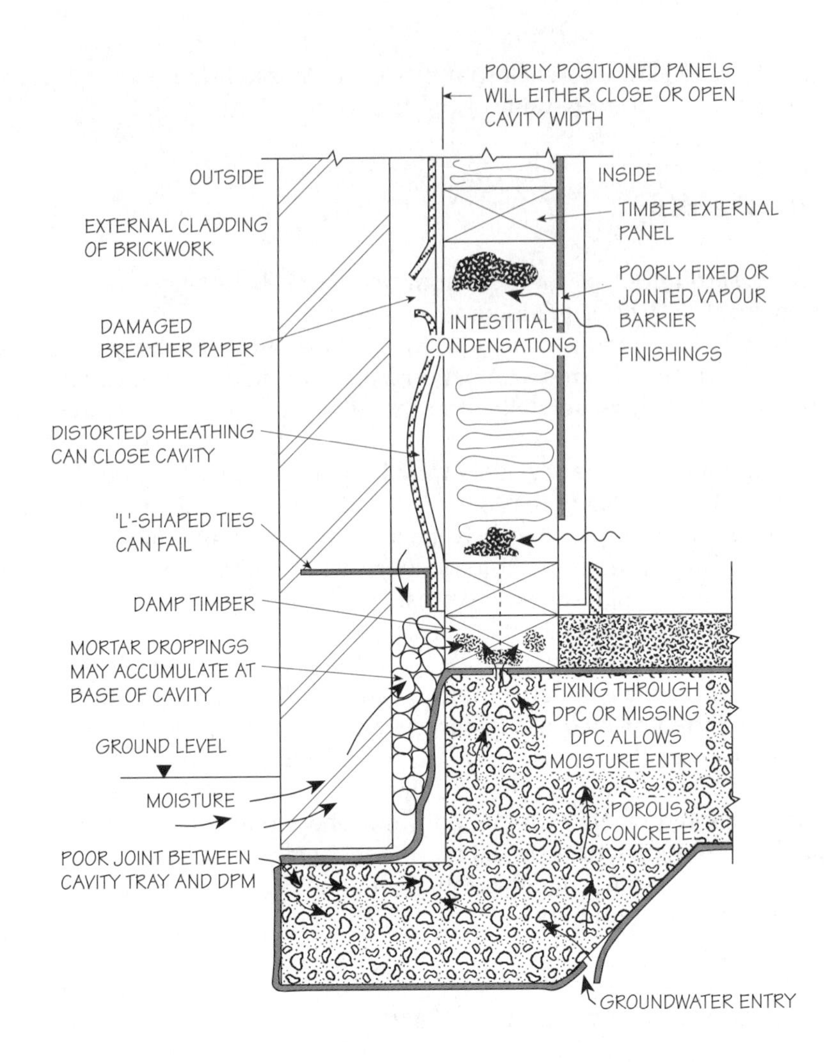


**Fig. 9.22.** Timber frame – general defects with external walls.

Any fixings for vertical timber cladding, tile hanging or rendering may puncture the breather paper. The omission of counter battens to provide a drainage and ventilation cavity may increase the risk of decay of the battens. Vertical battens are usually sufficient for horizontal shiplap boarding. Ferrous metal lathing used as a key for rendering can fail owing to corrosion. Vertical or horizontal timber cladding or rendering fixed directly to the timber panels may trap water between the boarding and the breather paper, increasing the risk of decay or dampness penetration.

The moisture movement of the timber cladding is likely to be significant. Boarding which is double-nailed or abuts masonry is unlikely to accommodate this movement.

Fixing felts rather than breather paper on the outside of timber panels may provide some resistance to rain penetration but can trap moisture within the timber panel.

The general trend to provide considerable thicknesses of insulation in timber panels accentuates the cold bridging effect when insulation is omitted or poorly fitted. This is more likely to occur at corners or other locations where fixing insulation is relatively difficult and heat flow is

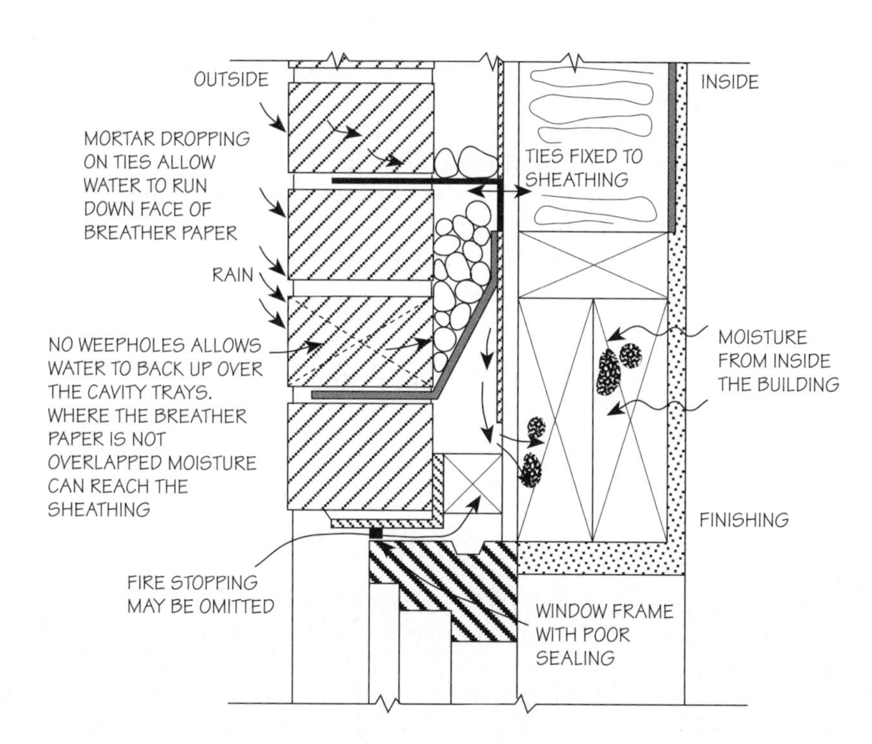


**Fig. 9.23.** Timber frame – general defects at window head.

concentrated. Surface condensation may occur on the internal finishes, which may be a symptom of missing insulation.

Omission of vapour barriers, or their incorrect positioning on the cold side of the insulation, can cause interstitial condensation to occur within the insulation. Since the breather paper will generally only allow a slow rate of moisture diffusion into the structural cavity and there is little ventilation between the insulation and the breather paper, there is a risk of decay of the timber panels.

The joints in the vapour barrier, commonly lapped, may allow moist air to enter the insulation. Any general cutting of the vapour barrier to accommodate services may provide additional points of entry. The general thickness of 500 gauge polythene vapour barriers is 0.127 mm and this can be readily damaged by on-site activities.

## Fire barriers and stopping

Cavity fire barriers can only be effective when they close the complete cavity. Therefore any failure of the ties to maintain the minimum cavity width may considerably reduce the effectiveness of cavity barriers. Cavity barriers, required at the eaves and between dwellings, made from timber may be subject to decay when DPCs between the timber and the outer leaf of a cavity wall are omitted. Fire stopping, required between external and separating walls in dwellings and where the cavity provides direct access to a structural element, may be provided by mineral fibre quilt. Unreinforced types can sag and distort and the quilt may provide a pathway for dampness.

Fire and smoke may be able to pass between the roof spaces of terraced or semi-detached dwellings. This can occur when the separating wall construction has not been extended to the underside of the roof covering and there is inadequate fire stopping under the roof covering.

There can be particular fire hazards associated with treating timber by spraying them with timber preservatives. Where the preservatives may soak insulation or fire-resistant quilt there is a considerable risk of auto-ignition. Specialist advice should be obtained.

### Revision notes

- Timber-framed buildings can be affected by the range of defects which affect timber, although they commonly appear at an advanced stage. Changes in the moisture content of the timber between storage on site and conditions in use can cause shrinkage, generally in the vertical direction, and corrosion of ferrous fixings. Factory-applied preservatives may be odourless and colourless, complicating identification.
- Inaccurately setting-out a timber frame can cause a reduction or increase in cavity width; this can increase the risk of rain penetration and reduce the bonding provided by the cavity wall ties. 'L'-shaped cavity ties may not provide structural integrity. Cavity fire barriers or stopping may be ineffective and breather paper can be punctured.
- A lack of adequate fixing of the binder rails may compromise the structural integrity of the frame. Sole plates can be inadequately fixed or penetrate the DPC. Any discontinuity between the DPM and the DPC may allow dampness to penetrate.
- The significant thicknesses of insulation in timber panels can accentuate any cold bridging and interstitial condensation. Thin vapour barriers can be punctured.

## Discussion topics

- Produce a checklist of items to be examined when inspecting a timber-framed building.
- Compare the defects associated with timber-framed and precast concrete industrialized buildings.
- Describe the movement characteristics of a timber-framed building which is clad externally with an external leaf of brickwork.
- Illustrate the possible routes of dampness penetration associated with a window opening and a difference in ground floor level across a separating wall, of a timber-framed building.
- Describe the function and possible failure mechanisms associated with the components of an external timber-framed wall panel.

- 'The use of timber as the only structural material in dwellings raises serious concerns with regard to their performance in fire.' Discuss.

---

## Further reading

BRE (1986) *External Walls: Brick Cladding to Timber Frame – The Need to Design for Differential Movement (Design)*, Defect Action Sheet 75, Building Research Establishment, HMSO.

BRE (1986) *External Walls: Brick Cladding to Timber Frame – How to Allow for Movement (Site)*, Defect Action Sheet 75, Building Research Establishment, HMSO.

BRE (1986) *External and Separating Walls: Cavity Barriers Against Fire – Installation (Site)*, Defect Action Sheet 75, Building Research Establishment, HMSO.

BRE (1990) *Sound Insulation of Lightweight Dwellings*, Digest 347, Building Research Establishment, HMSO.

BRE (1991) *Design of Timber Floors to Prevent Decay*, Digest 364, Building Research Establishment, HMSO.

BRE (1991) *Bracing Trussed Rafter Roofs*, Good Building Guide 8, Building Research Establishment, HMSO.

BRE (1993) *Supplementary Guidance for Assessment of Timber Framed Houses: Part 1, Examination*, Good Building Guide 11, Building Research Establishment, HMSO.

BRE (1993) *Supplementary Guidance for Assessment of Timber Framed Houses: Part 2, Interpretation*, Good Building Guide 12, Building Research Establishment, HMSO.

Cook, G.K. and Hinks, A.J. (1992) *Appraising Building Defects: Perspectives on Stability and Hygrothermal Performance*, Longman Scientific & Technical, London.

Council of Forest Industries of Canada (1981) (revised 1992) *Timber Frame House Construction: Check it Out.*

Desch, H.E. (1981) *Timber: its Structure, Properties and Utilisation*, 6th edn (revised by J.M. Dinwoodie), Macmillan Education, London.

Richardson, B.A. (1991) *Defects and Deterioration in Buildings*, E.&F.N. Spon, London.

TRADA (1992) *Surveys of Timber Framed Houses*, Wood Information Sheet 0/10. Timber Research and Development Association.

# 9.5 Cladding defects generally

## Learning objectives

You should understand:

- the range of defect mechanisms which can affect cladding;
- the difficulty of diagnosis (and repair) with this expensive and critical element of buildings.

Cladding is distinguished from conventional solid or cavity walling on the basis of its functions. There are usually obvious constructional differences although these may be concealed by the exterior design appearance of the building.

There are problems of definition over the transition from vertical or near-vertical cladding to roofing (consider, for example, Charing Cross Station). The distinction in terms of occurrence and identification of apparent defects is important, however, since roof glazing can present special problems it is better studied individually. Further distinction may be made within the category of claddings, of loadbearing and non-loadbearing elements.

The broad classification of loadbearing and non-loadbearing can easily overlook some of the most important cladding roles and most likely sources of defects. These are the resistance to imposed wind loads without excessive deflection or movement of the cladding and its framing, whilst maintaining its critical weather-tightness properties, and also the transfer of these dynamic loads, which may be a positive or negative form and cycling rapidly, through the fixings to the building.

Very rarely does the modern external cladding skin contribute to the loadbearing capacity of the structure, and it may be useful to distinguish between this type of loading and the supporting of environmental forces and self-weight alone. Indeed, the failure of claddings under acquired loading confirms this distinction.

There has, however, been a change in the understanding of the term 'cladding' and whilst failures attract media attention the actual number of incidents is relatively low taken in the context of the amount of cladding used. Indeed, there appears to be a relatively good serviceability record. The cost implications of failure in use are extreme, however.

*The Technology of Building Defects.* Dr John Hinks and Dr Geoff Cook.
Published in 1997 by E & FN Spon, 2–6 Boundary Row, London SE1 6HN, UK. ISBN 0 419 19770 2

Claddings can fall, and have fallen, off buildings where the fixing detailing has been deficient (or has deteriorated in use owing to corrosion or fatigue). It is obviously important that the fixings are durable, but also that they can accommodate differential movement between the envelope and building frame, and also that any structural secondary framing is adequate for transferring these forces.

There is a degree of similarity in the types of shortcoming occurring with the variety of claddings. Most are related to their relationship with the environment. The real problems occur where the cladding interfaces with the framing or the surrounding building.

The significance of faults in the stability of cladding lies firstly with the difficulty of identification because of the concealed nature of fixings, and the impracticality often of inspection. Secondly, there is the possibility that they will emerge catastrophically. The obvious implications of falling cladding panels are sufficient reason alone for concern and attention to the avoidance and identification of deficiency. This has not gone unnoticed by the professional institutions and the standing committee on structural safety has reported on the problems of cladding safety. Sub judice rules or agreements barring release of information identifiable with particular building failures form recurrent barriers to learning from constructional failure. In the USA information about failures is frequently more readily available in the public domain and the benefits of hindsight can be taken.

The majority of the failures in claddings fortunately fall short of affecting the structural stability. The frequent problems lie with serviceability and the envelope, a primary function with the non-loadbearing claddings. The effects on the everyday use of the building and its internal climate are more obvious, and the enormous costs associated with the modification or replacement of the cladding are a key reason for the importance of good serviceability, or fitness for purpose.

This distinction is clear in the standard tests and much effort is concentrated on producing water- and wind-tight claddings. The difficulty in producing this repeatably and reliably on site is twofold. Firstly there is the translation of theoretical design into practical reality, and secondly the problems of site workmanship quality and its supervision. There is ample evidence of cladding defects from the 1960s UK stock. The problems with this are sufficient to encourage testing facilities to be used, usually prior to use or completion of the building. It is not unknown for testing and the performance of the materials in practice not to relate directly. The tests are short term and do not necessarily reflect the lifetime performance. Tests on in-situ work only examine the potential that can be achieved. This need not be through any avoidable shortcomings in the test, rather the testing of an in-situ panel leaves variability on site an unknown factor. There may also be a disparity between the mock-up and the full-scale applications.

In the United States the criteria for satisfactory performance may be more extensive than for the UK, particularly in the area of seismic activity. Here the testing of panels frequently requires racking loads to be imposed on the tested panelling, followed by retesting to their normal criteria. Note that for impact testing a cracking of a glass panel may not be classed as a failure.

The ingress of water not only involves moisture but also frost and snow. The wind permeability of the structure is an important performance criterion and tests make no distinction between air ingress around the whole frame or at individual locations.

Damp control extends to rainscreen cladding also, where the provision of ventilation into the building must avoid directly crossing the external cavity, otherwise damp air is pulled into the building and introduces the risk of condensation.

Care also has to be taken with the mixing of timber and aluminium as an external shield to the building frame. This can easily provide degradation and damage to the wood as the high conductivity of the aluminium lowers the temperature within the wood profile locally. Where the aluminium covers the wood completely there is a more uniform drop in temperature and the wood components seem to respond better.

Connections of the framing of the cladding structure to the substrate must be made with care. If the connections between the frame (subjected to the warmth and dampness of the interior environment) and the substrate are made on the cold side of the insulation layer there is a temperature drop within the connector. In effect the connection forms a cold bridge and in the absence of an effective vapour barrier condensation can occur.

Differential movement has also been one of the principal cladding problems, either in the context of relative movement between the external envelope and the supporting substructure or within multilayered claddings. Diurnal and seasonal thermal movements within the cladding and moisture-related movements in the permeable claddings compound the dynamic relationship between the supporting structure and the envelope.

The extent of movement occurring within lightweight, thermally insulated claddings can be extreme, especially with dark facings which can absorb a great deal of heat. Where there are long runs of dark material with little provision for expansion the corner details may fail as the expansion of the coverings meets and stresses conflict at a change in direction. The movement of such panelling is usually relatively unrestricted compared with the tying-in effects occurring from self-weight in conventional or heavyweight claddings.

Frequently the manufacturer and site assembly staff are blamed for faults in the workmanship of the systems. Indeed there can be severe environmental problems on-site, against pressure for rapid completion.

There is a factor of buildability of the design also. Poor detailing that makes correct assembly on site or in the factory impractical is also a workmanship fault.

Tolerances have been a problem in the past, particularly with the concrete systems of the 1960s. Control and coordination of tolerance within a closed system of cladding should be readily attainable; however, the coordination with the building frame, particularly in-situ concrete frames, is more difficult. The manufacturing tolerances achieved in the high-quality modern curtain walling systems can outstrip those attainable in high-quality in-situ concreting, simply because of the nature of the materials involved.

There is also a particular need for high-quality supervision where the detailing is unconventional, since genuine misunderstandings or apparently superficial changes to the detail on site can cause failure.

The absence of thermal breaks on the early curtain walled systems led to severe problems with surface condensation on the inside of the structure. Developments have not been without their problems either. Recommendations are now that thermal breaks be installed, as near to the face of the cladding as practicable (to minimize the infrastructural condensation problem).

Difficulties or impracticalities in carrying out planned maintenance during the interim may shorten the life of the cladding.

The position and reliability of vapour barriers are critical. Many specifications will be calling for sheet metal vapour barriers rather than polythene for durability's sake. Moisture control is one of the most critical actors for serviceability and durability of the cladding structure. It is frequently associated with the premature failure of conventionally attached claddings.

There have also been problems with the staining and deterioration in polluted environments of the anodized aluminium coatings.

On-site conditions are a potential cause of failure. In addition to the chronic citing of supervision as the cause there are also possible problems with the weather and general environment in the use of sensitive sealants.

## Revision notes

- Functional differences with cladding and conventional walling, in addition to constructional differences.
- Non-loadbearing and loadbearing cladding distinctions affect nature and significance of defects.
- Permeable and non-permeable cladding.
- Similarity in general problems: interfaces with framing and cladding, and/or framing and building; damp penetration; differential movement; workmanship; tolerances; thermal breakage/cold bridging; position and reliability of vapour barriers; staining; deterioration of coatings and sealants.
- Difficulty of testing cladding.

## Discussion topic

- Compare and contrast the types of defect occurring in claddings with conventional masonry walling, and comment on the similarities in mechanism. Discuss the differences in consequence of deficiency in the various forms of wall.

## Further reading

AWA *Curtain Walling and Large Composite Windows: Notes for Guidance*, 2nd edn, Aluminium Window Association.

Battrick, G.N. (1989) Proto-type testing. *Proc. Conf. Cladding: Current Practice in Specification, Design and Construction*, 7 November.

Brooks, A. (1993) New standard for curtain walling. *Architect's Journal*, 14 April, p. 39

CWCT *Standard and Guide to Good Practice for Curtain Walling*, Centre for Window and Cladding Technology.

Endean, K. (1995) *Investigating Rainwater Penetration of Modern Buildings*, Gower.

ISE (1995) *Aspects of Cladding*, Institution of Structural Engineers.

Hugentobler, P. (1989) Design and development. *Proc. Conf. Cladding: Current Practice in Specification, Design and Construction*, London, 7 November.

Muschenheim, A. and Burns, S.J. (1989) Specification 1: the specification of cladding at Skidmore, Owings and Merrill. *Proc. Conf. Cladding: Current Practice in Specification, Design and Construction*, 7 November.

Tietz, S.B. (1989) Cladding: Current Practice in Specification and Design. *Structural Engineer* **67** (21), 389.

White, R.F. (1989) The Benefits of Prefabrications. Lessons from Cladding. *Proc. Conf. Cladding: Current Practice in Specification, Design and Construction*, London, 7 November.

# 9.6 Permeable claddings

## Learning objectives

- You should be aware of the behaviour of permeable materials and the effect this has on deficiencies in use as cladding.
- The difficulty of assessment is an important issue.

## Stone

With the traditional heavyweight forms of cladding, such as supported Portland stone slabs of 100 mm thickness and granites and marbles commonly cut to 40 mm, and also 200 mm brickwork, the self-weight of the cladding makes a significant contribution to the stability of the panels. Their inertia helps them resist the rapidly changing forces, and their tying back with fixings was largely unremarkable. However, there was a period where the fixing of all four sides of a panel was common practice. This prevents relief of secondary stresses arising from thermal or moisture movement in the panels, and damage occurs.

There have been reports of failures or deficiency in the envelope reported where the permeable skins and the concealed steel structure have been constructed in direct contact, resulting in rusting of the steel frame, and concerns over the position of weatherline and protection of the structure were sometimes overlooked. The problem with permeability in brick cladding is largely dependent on the mortar.

The use of these materials enjoyed particular popularity in the 1980s, although cost and weight implications for the frame and foundation design, and also maximizing the rentable space, led to the replacement of the traditional thick panels with thin, sometimes laminated forms of insulated wall construction. The general concept emerged as a unitized or panel construction.

Whilst retaining the traditional and monumental appearance, the connections to the building and the relationship with the cladding material changed. In particular the use of rigid panelling with flexible (steel) structures was highlighted in the USA as requiring careful design. There is at least one recorded instance in the USA where complete recladding has been required following distortion in thin-sliced stone panels.

Consequently, it now contrasts less sharply with the lightweight frame and glazed panel systems and the emerging structural silicone glazing. In

*The Technology of Building Defects.* Dr John Hinks and Dr Geoff Cook.
Published in 1997 by E & FN Spon, 2–6 Boundary Row, London SE1 6HN, UK. ISBN 0 419 19770 2

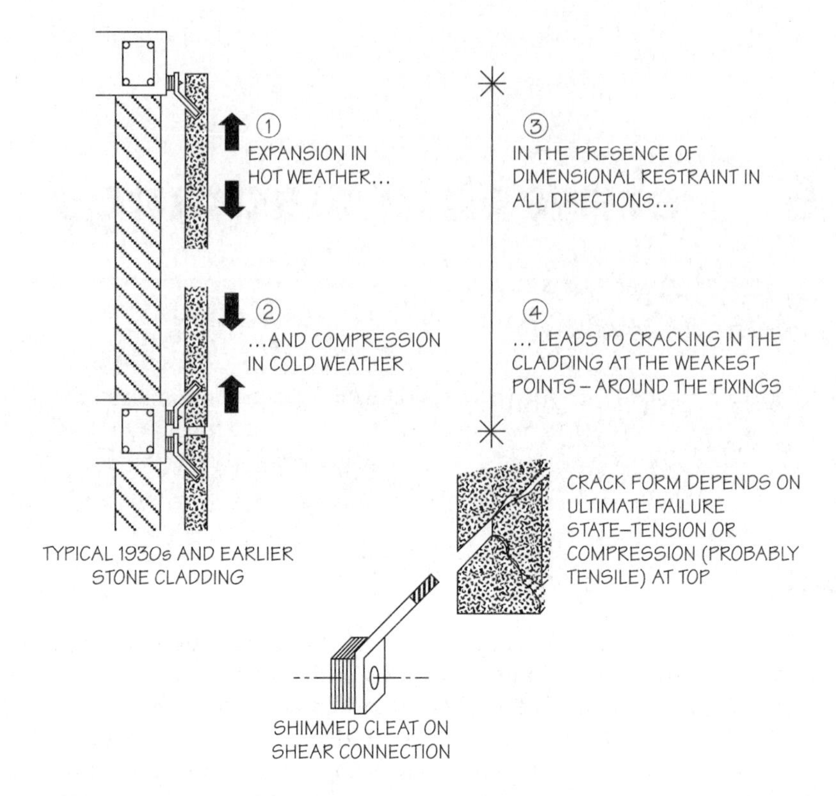


**Fig. 9.24.** Failure under movement stresses within restrained stone cladding.

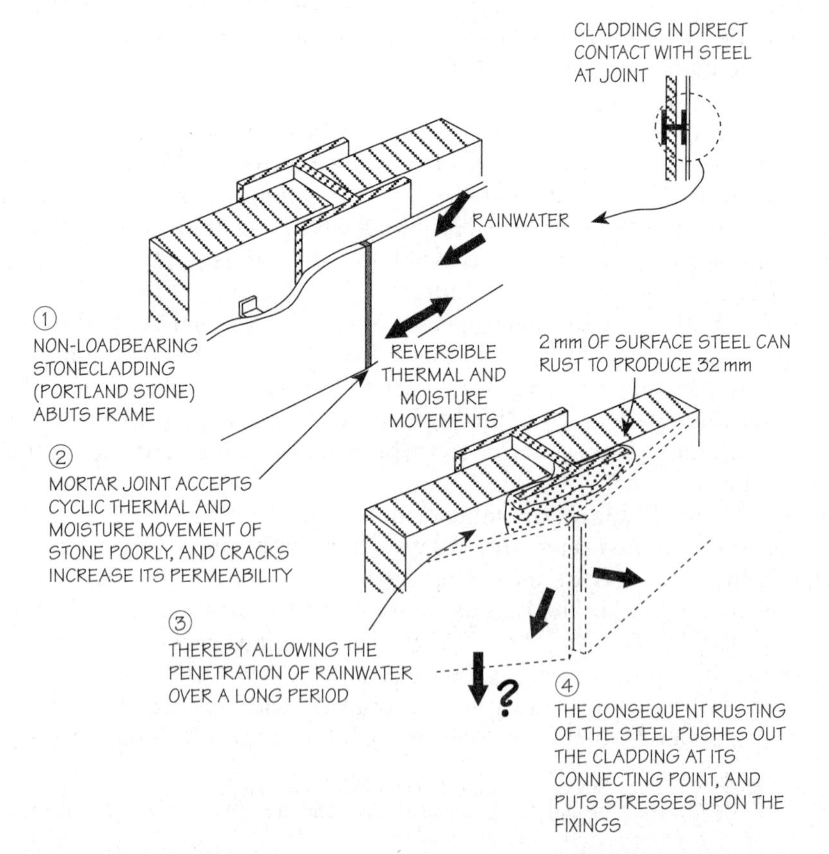


**Fig. 9.25.** Schematic diagram for cladding failure due to framework corrosion.

particular the position of the weatherline in the wall construction has been pushed outwards.

## Precast concrete cladding

The precast concrete cladding symbolic of the systems boom is highly variable. The tying to the structure is critical with these forms of construction to prevent problems occurring with differential movement as the frame initially shrinks and experiences long-term creep; 3 mm in 3 m is possible. The possibility exists of acquired loading as the vertical shortening of the frame transfers loads onto the cladding. Symptomatic failures include bowing and cracking as the outer skin is tensioned, possibly culminating in the complete failure of panels as the nibs connecting them with the frame shear off under loads they were not designed for.

Early glass-reinforced cement (GRC) panels, constructed using alkali-resistant glass-fibre, may suffer from cracking, the dark-coloured panels being the more susceptible. Surface crazing is possible, as with concrete panelling. Possible causes include design limitations, excessive restraint and/or poor manufacture. It is important to appreciate that GRC undergoes greater moisture movement than in comparable precast concrete panels.

## Brick cladding

The brick-clad structures of the 1970s also suffer problems with differential movement. A common cause of this is the counter-stress of in-situ concrete frames undergoing drying shrinkage and clay brickwork expanding.

Where there is inadequate provision of expansion jointing, the brick panels may buckle and dislodge. This can produce failure at the weakest and thinnest parts of the brick cladding, the slips at floor levels. This has been confirmed as a significant source of problems in the past, and to a lesser extent the quality of fixing of slips is a source of problems.

An alternative mode of failure can occur with the concrete nibs supporting the brickwork being pulled out from the remainder of the frame by the distorting brickwork as it is compressed. Where the expansion of the outer leaf is significantly involved, an inner concrete block leaf may show few or no signs of distress.

Joints and jointing technology have been the central issue in failure over the last twenty years in modern claddings. Vast improvements have been made. The traditional mortar joint as used for the brickwork cladding panels and as connections in the construction of stone-clad structures has an average life expectancy of probably about 20 years.

### Revision notes

- Stone, concrete, brickwork.
- Problems with adjacent framing where the permeable cladding produces a route for damp penetration.

- Problems with materials and components themselves.
- Thin stone claddings have potential movement and distortion problems. Fixing detailing and movement accommodation are very important.
- Precast concrete cladding suffers general problems of bowing cracking, staining and differential movement with the frame. As with stone claddings, the detailing for movement is important. Nib failure can be problematic.
- Brick cladding moves differently from the frame and space for differential movement was not recognized in early uses. Brick slips can also be problematic.
- Glass-reinforced cement panels may exhibit surface crazing or cracking due to a range of possible causes – poor design, poor manufacture and or excessive restraint.

## ■ Discussion topic

- Compare and contrast the general defects associated with permeable cladding and conventional permeable walling.

## Further reading

ACA *Fixings for Precast Concrete Panels, Data Sheet 1*, Architectural Cladding Association.

BDA (1986) *Brick Cladding to Steel Framed Buildings*, Brick Development Association, in conjunction with British Steel Corporation.

BRE (1979) *Wall Cladding: Designing to Minimise Defects Due to Inaccuracies and Movements*, Digest 223, Building Research Establishment.

BRE (1980) *Fixings for Non-loadbearing Precast Concrete Cladding Panels*. Digest 235, Building Research Establishment.

BRE (1982) *Reinforced Concrete Framed Flats – Repair of Disrupted Brick Cladding*, Defect Action Sheet DAS 2/82, Building Research Establishment.

BRE (1985) *Wall Cladding Defects and Their Diagnosis*, Digest 217, Building Research Establishment.

GRCA. *Guide to Fixings for Glassfibre Reinforced Cement Cladding*, Glassfibre Reinforced Cement Association.

GRCA. *Guide to Specification for Glassfibre Reinforced Cement*, Glassfibre Reinforced Cement Association.

Moore, J.F.A. (1984) *Use of Glass Reinforced Cement in Cladding Panels* Report 49, Building Research Establishment.

# 9.7 Impermeable claddings

**Learning objectives**

- You should understand the particular problems with impermeable claddings such as rainwater handling and water penetration generally.
- There can also be specific problems with thermal movement.

The emergence of the lightweight impermeable skin has produced a distinct change in the approach to achieving performance criteria. The structural criticality remains, but imposed over this is the need to allow for physically significant and frequently differential movement within the envelope. The well-publicized howling failures in the United States are still referred to: the cracking or loss of panels as frame and infill moved differentially with inadequate allowance. The advancements in the technology have been significant and the cladding systems emerge from being very large windows, through the early single-glazed and front-sealed forms of patent glazing, to dynamically responsive skins.

## Proprietary glazing faults

Proprietary glazing systems can range from basic sloping glazing with transoms omitted, and liable to leakage and staining in the capillary joint this forms, to adaptations of curtain walling systems. Exposed transom details in inclined roof applications provide a barrier to simple water run-off, and in such instances the quality of seal will be critical and continuously tested. Water leakage problems have been reported with the early forms that use poorly designed or assembled neoprene sealing strips. The development of the hermetically sealed double glazed unit and thermally broken frame have reduced the problem with condensation experienced by the early designs.

## Curtain walling

The curtain wall must be capable of supporting the stresses induced by self-weight and the wind loads. The wind may produce positive and negative

*The Technology of Building Defects.* Dr John Hinks and Dr Geoff Cook.
Published in 1997 by E & FN Spon, 2–6 Boundary Row, London SE1 6HN, UK. ISBN 0 419 19770 2

forces, which create compressive, tensile, shear and bending stresses in the glazing and frame.

The deflection limit for storey-height framing under the design wind loading should be 1/200 clear span, although the limit for insulated glass is 1/175 and framing to plain single glazing is suggested not to exceed 1/125.

## Glass and frame relationship

There have been reports of many failures with glass, because of strength and inclusion in the façade. It should obviously be necessary to have sufficient clearance around the enclosed glazing to allow for the differential expansion and contraction of the metal framing. Aluminium can have a coefficient of thermal movement exceeding twice that of plain glass, and a gap of about 30 mm is necessary for modest-sized panels. The dark-coloured and solar-absorbent types of glass will have thermal expansion coefficients approaching nearer that of aluminium, and greater movement must be allowed for. It is common to assume a maximum design temperature of $+70°$ C although it is possible for $+90°$ C to occur on the thermally modified glazing. Extreme movement occurs with insulated rear faces, producing heat retention in the panel during hot weather and temperatures similar to less than outside on winter nights. Surface temperatures of $-25°$ C are common, and night sky radiative temperatures assume $-40°$ C temperatures. With a possible temperature extreme scale of 130° the potential physical movement is clear (approximately 20 mm/m run difference between glass and aluminium). The thermal expansion coefficient of glass is $9 \times 10^{-6}$ per °C, and that of aluminium $24 \times 10^{-6}$ per °C.)

The alternatives to correct tolerances in the cladding systems are loss of retention of the panels or crushing of the glazing, normally producing a single crack. These have been experienced in the USA and the lessons apparently now learnt.

Thermal movements within the framing will be physically significant, more so in the mullions because of their greater dimensions. The sealant or gasket movement joints must obviously be sufficient to allow expansion and contraction of the framing without damage. The thermal conditions at the time of installation of the gaskets will also be an important feature of their suitability for accommodating the extremes of movement. The transfer of the movement in the curtain wall framing must be carried through to the structural frame.

The detailing of the connection between the glass and the frame is important for the control of water ingress. The impermeable wallings shed all rainwater, unlike the traditional permeable wallings which only discharge that which is not absorbed. This creates tremendous pressure on the quality and longevity of waterproofing, particularly horizontal seals at the transoms over which most of the rainwater discharges. This was identified as a real problem with the early curtain walling systems which were commonly front-sealed with a mastic sealant. Under prolonged exposure to sunlight the sealants invariably dried out and cracked. This produced direct or capillary leakage paths for wind-driven rainwater to enter the frame structure.

**Fig. 9.26.** Lightweight cladding – intolerance of differential movement between panel and frame. (K. Bright.)

A development to avoid the problems of the water becoming trapped, and an acquiescence in the ingression problem, were the drained and vented systems. These had an improved neoprene or silicone gasket form of seal, and the reduced amounts of water still pumped through the seal under gusting were fed out through drainage slots in the curtain wall framing. Joints and sealants will be one of the main areas of the curtain walling systems which will probably have a shorter life span than the remainder of the framing. They will require frequent inspection (rarely simple). Where the expansion joints in the framing have been remade the drainage slots may be blocked.

## The relationship between defects in frame and structure

The connection of the curtain wall framing to the structural frame of the building is also an important source of potential defects. With concrete frames there is a risk while the walling and frame are exposed to the weather. Aluminium and its oxides are attacked by the alkaline environment in cement-based products, and the hydroxide salts dissolved in any rain-water running off the concrete frame or present during curing can damage the anodizing and the base metal.

The attachment of aluminium framing to steel structures should be done with insulating washers separating the two metals. Direct contact of the two dissimilar metals will set up a galvanic cell and corrosion of the aluminium will occur.

On the outside of the building, the exposed aluminium framing common in older curtain walling systems is prone to degradation and staining from environmental pollution. Staining of the anodized coating may be removable with care, although use of aggressive cleaners will strip the anodizing also.

## Shrinkage of the structural frame

The shrinkage of the concrete-framed structure can still produce problems. Where the lower floors shrink or deflect, the loading is transferred to the curtain walling. Once the tolerance of movement in the joints is exhausted, the loading racks the glass panels. Alternatively, at the roof level the relative rigidity of the parapet wall will limit deflection of the roof slab only. The cumulative effect of the floors sagging below can cause falling-out of the glass panels.

## Glass-reinforced plastic cladding

Glass-reinforced plastic claddings may suffer from weathering deterioration, in particular degradation as a result of exposure to ultraviolet light. Dirt collection upon rainwater washing will stain the surface. Because of the relatively high thermal movement coefficient, it is important to check movement joints.

## Structural silicone

There have been very few structural silicone failures reported so far, although this is still a relatively new technique. In Australia there is experience of the use of structural silicone over ten years, to 50 storeys. Green (1989) reports a failure rate in the USA of below 0.1%, with no

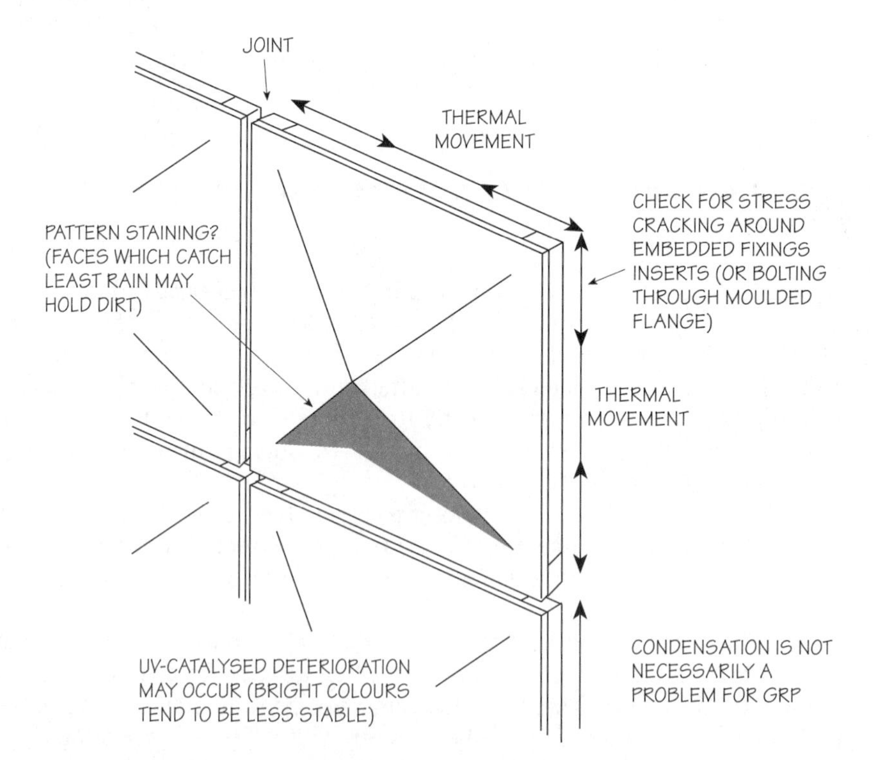


**Fig. 9.27.** Check details for glass reinforced plastic cladding panels.

instances of glass falling off and all failures occurring within the first three years of life. The few problems are reported to have been traced to the adhesion between the silicone tensile bead and the metal substrate (mullion or transom) rather than between the glass and the silicone. Reviewing curtain wall engineering, Hunton and Martin (1989) make the following sobering comment about the Australian market: 'there is strong evidence that the concept of structural silicone glazing continues to be marketed without sufficient engineering review.'

Cleanliness during assembly is of paramount importance, and is particularly difficult to achieve on site. The mechanism of deterioration of the joint is likely to be related to the response of the silicone to deflection of the joint under alternate positive and negative wind-loading pressures. It is worth noting that an initial contraction in the silicone of about 2–3% has been reported. Any voiding in the silicone or damage to the surface of the exposed jointing should be avoided.

Obviously, wherever a component of a building is critically dependent on a single line of support then planned inspection is a must. In these cases the quality of assembly and the adhesion and durability of the silicone are critical. Many of the systems, however, have a secondary support in the form of concealed or disguised corner cramps.

In the UK, gaskets followed the early gunned sealants, but had characteristic faults of springing out of the joint or being pulled out under cleaning operations.

**Revision notes**

- Curtain walling, proprietary glazing.
- Lightweight impermeable claddings can have rainwater handling problems at joints.
- Water entrapment: can be difficult to identify the source.
- Problems at the interface between glazing and frame, also with dark colours being liable to high thermal gain and movement.
- Thermal movement is a principal problem.
- Problems also arise at the frame-to-structure interface; relative movement problems may be reversible (cyclic) or irreversible.
- Other faults include staining, glazing dislocation and sealant failure.
- Structural silicone is highly dependent on workmanship and site conditions for durability. Early silicone-only fixings were superseded by systems with secondary support.

## ■ Discussion topic

- Compare and contrast the mechanisms and symptoms of defects in permeable and impermeable claddings.

## Further reading


BRE (1974) *Reinforced Plastics Cladding Panels* Digest 161, Building Research Establishment.

Campbell, J. (1989) Cladding specification. *Proc. Conf. Cladding: Current Practice in Specification, Design and Construction*, London, 7 November.

Gilder, P.J. (1989) Lessons from cladding. *Proc. Conf. Cladding: Current Practice in Specification, Design and Construction*, London, 7 November.

Green, S. (1989) Choice of system – American current practice *Proc. Conf. Cladding: Current Practice in Specification, Design and Construction*, London, 7 November.

Hunton, D.A.T. and Martin, O. (1989) Curtain wall engineering. *Proc. Conf. Cladding: Current Practice in Specification, Design and Construction*, London, 7 November.

Josey, B. (1992) Paneful inclusions. *Building*, 2 October, pp. 60, 61.

Townsend, J. and Bennett, D. (1985) Patent glazing. *Architect's Journal* **182**(33), 50–51.

# 9.8 Problems with lightweight metal cladding

## Learning objectives

- You should be able to understand the causes and effects of a range of defects which can occur with lightweight metal cladding.
- In addition to the influence of corrosion you will be made aware of the failure mechanisms of joints and the role of external forces.

## Defects generally

Lightweight metal non-loadbearing cladding carries and transmits live loads in the form of wind loads. These loads may be significant owing to the relatively large surface areas of cladding and can distort or blow off the cladding. Any distortion of the supporting framework may also impose loads on the cladding.

It is essential that any cladding should maintain its weatherproofing integrity whilst accommodating the movements of the structure. Where this has not occurred the cladding may become defective.

There is a general difficulty in carrying out inspections of cladding. The lack of access to the rear of claddings can be a particular problem. This can mean that any defects may not be identified until symptoms are obvious.

Metal claddings are impermeable, causing water to flow rapidly into the roof drainage system. This may fail where carrying capacity is reduced by inadequate size, poor falls or debris.

## Corrosion

Drilling and impact damage can allow corrosion of ferrous metal cladding sheets to occur. This will be accelerated by the differential oxygenation effect as corrosion moves under the protective layer away from the initial site of the damage. The overall breakdown of the protective coating to steel cladding can occur. This may be due to corrosion of the steel or a failure of

*The Technology of Building Defects.* Dr John Hinks and Dr Geoff Cook.
Published in 1997 by E & FN Spon, 2–6 Boundary Row, London SE1 6HN, UK. ISBN 0 419 19770 2

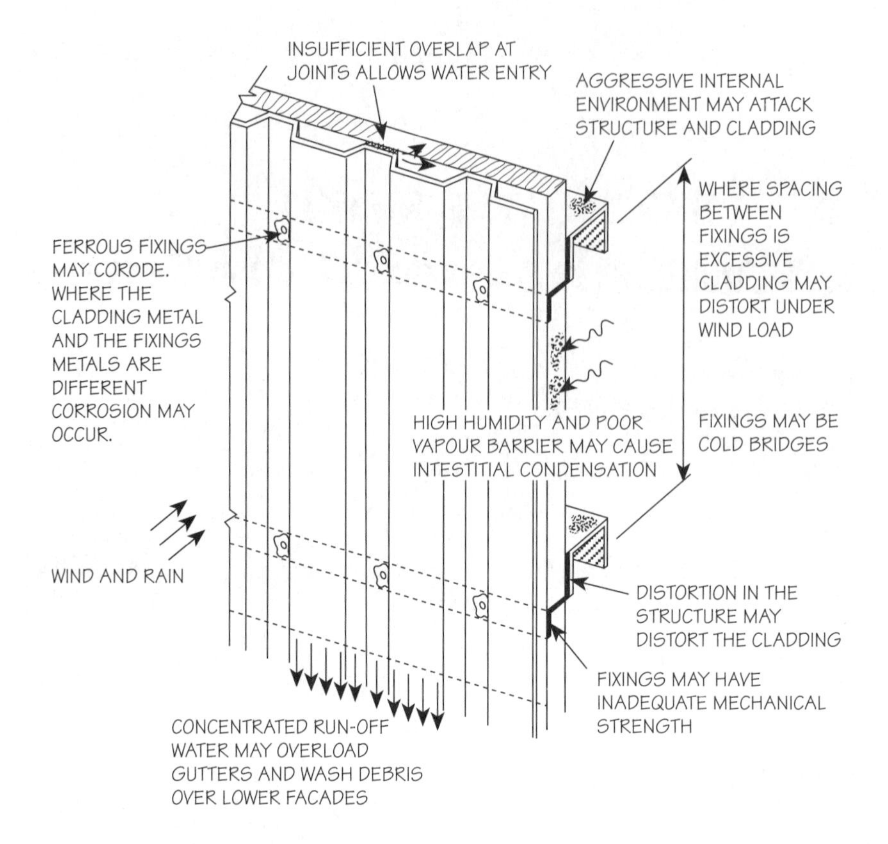


**Fig. 9.28.** General defects in lightweight metal cladding.

the bond between the coating and the cladding. Fixings are also exposed to atmospheric corrosion.

## Joints and fixings

Inadequate fixings may provide insufficient support for the sheets. Fixings in the trough of profiled sheets can leak. Snow a greater depth than the profile can cause leakage, not only in fixings through the crown of the profile but also by capillary action through lapped joints. The accommodation of movement and the provision of a secure fixing may be a contradiction. With profiled sheets the movement may be accommodated by the relatively small dimensions and changes of direction of the profile. Longitudinal thermal movement, which may be significant for dark-coloured sheets facing the sun, is commonly accommodated by unfixed, lapped joints. Rain may be blown through these joints where the lap is insufficient. Many cladding systems are based on movement accommodation being provided by the supporting structure. Where there are material differences then there can also be movement differences.

The movement of the sheets at joints and fixings due to wind or thermal movement can wear away any protective coating and allow corrosion sites to form. The inner sides of cladding sheets which are exposed to a corrosive environment in the building can become coated with hygroscopic salts. These can absorb moisture from the air, which could accelerate any corrosion of the sheets. Corrosion can also occur on aluminium cladding sheets.

## Differential movement

Bowing of sandwich construction cladding can be caused by differential thermal movement. The external metal cladding layer may become detached from the insulating core when exposed to high temperatures. Where there are long runs of dark cladding material with little provision for expansion the corner details may fail as the expansion of the cladding abuts the structure.

Permanent glazing can fail owing to differential movement of the glass and frame. Metals have a higher coefficient of thermal expansion than glass.

## Weathering

Weather resistance is usually afforded by joint overlaps which normally assume that water runs downhill. The amount of rainwater flowing down the face of impermeable metal cladding can increase the risk of water entering the joints. Wind action can drive water upwards through lapped joints. Anywhere that water can accumulate, capillary action through adjacent gaps in the cladding may also occur. In the case of sandwich construction, where a layer of insulation material is incorporated, then considerable quantities of water may be absorbed. Damp patches appearing on the underside of the panels, perhaps accompanied by sagging, are obvious symptoms.

Where aluminium sections are incorporated these are commonly coated for decorative purposes. Paint films can fail where the aluminium oxide layer disrupts adhesion. Acrylic coatings which are applied electrophoretically can provide adequate coverage, although they can fail where surface preparation of the aluminium is poor. The general durability of acrylic coatings is affected by a tendency for the surface to degrade because of weathering. Polyester powder coatings have largely superseded the acrylic coatings although they can still suffer from impact damage. Any on-site activities which break the protective coating may create locations for deterioration.

## Distortion

Since the panels tend to be thin any uneven fixing or movement of the supporting structure may cause distortion of the cladding panels. Wind loads may cause poorly secured panels to vibrate and even become permanently distorted.

Although there are differences concerning the definition of cladding, which is generally vertical and roofing, which is generally not so vertical, both can fail in similar ways. Some lightweight metal cladding systems are integral with the roof sheeting and adopt the same profile. Roof sheeting can be positively overloaded by snow and negatively overloaded by wind where roof pitches are low.

## Revision notes

- The weatherproofing integrity of any cladding may fail where it cannot accommodate the movements of the structure and any wind loading. Metals can corrode, particularly in aggressive environments, and protective coatings can fail.
- A general lack of access to the rear of claddings, or the inner cavity of sandwich construction cladding, can mean that defects may not be identified until symptoms are obvious.
- High water run-off rates from metal cladding can overload guttering and increase the risk of water entering the joints. Fixings in the trough of profiled sheets can leak.

## Discussion topics

- 'A range of defects in thin metal claddings is inevitable since the cladding should accommodate thermal movement and be securely fixed.' Discuss.
- Describe a checklist of features to examine when inspecting a large area of thin metal cladding.
- Explain how sandwich thin metal cladding, which contains an inner core of insulating material, may distort.
- Compare the role of corrosion in three defects associated with thin metal cladding.

## Further reading

BRE (1978) *Wall Cladding Defects and their Diagnosis*, Digest 217, Building Research Establishment, HMSO.

BRE (1979) *Wall Cladding: Designing to Minimise Defects Due to Inaccuracies and Movements*, Digest 223, Building Research Establishment, HMSO.

BRE (1990) *The Assessment of Windloads: Parts 1 to 8*, Digest 346, Building Research Establishment, HMSO.

Cook, G.K. and Hinks, A.J. (1992) *Appraising Building Defects: Perspectives on Stability and Hygrothernal Performance*, Longman Scientific & Technical, London.

PSA (1989) *Defects in Buildings*, HMSO.

Ransom, W.H. (1981) *Building Failures: Diagnosis and Avoidance*, 2nd edn, E. & F.N. Spon, London.

Richardson, B.A. (1991) *Defects and Deterioration in Buildings*, E. & F.N. Spon, London.

# 9.9 Solid brick walls

## Learning objectives

- You should be able to identify the factors which influence the deficiency of solid walls; this will also include the role of door and window openings.
- You will be made aware of the inherent defects associated with the use of bricks and stone to resist moisture, particularly in exposed situations.

## Solid walls – clay brick

Rain may pass through a solid wall through the small cracks within the mortar, or between mortar and brick. Exposed locations are particularly vulnerable. The perpends are particularly reliant on adequate workmanship to be effectively filled with mortar and where this does not occur alternative routes for rain penetration are provided. Poor-quality mortar may reduce the weather resistance of large areas of solid wall. Recessed pointing may provide ledges and crevices where rain can accumulate. This is then more

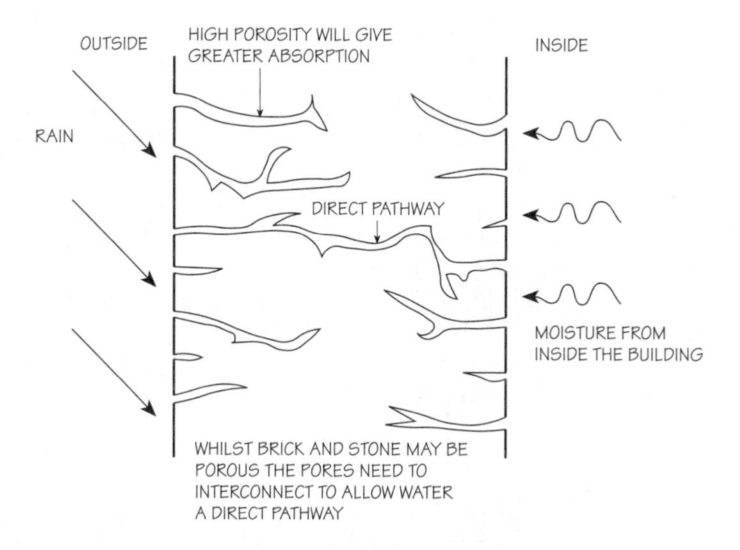


**Fig. 9.29.** Porosity problems with clay bricks.

*The Technology of Building Defects.* Dr John Hinks and Dr Geoff Cook.
Published in 1997 by E & FN Spon, 2–6 Boundary Row, London SE1 6HN, UK. ISBN 0 419 19770 2

likely to penetrate the wall and the saturated ledges are more liable to frost attack.

Since bricks are porous, except under very dry conditions the wall can contain moisture. Where water-repellent external surface treatments are applied under these conditions, the water within the wall becomes trapped. The evaporation of water from the external wall surface can be substantially reduced. Where dissolved salts would normally produce efflorescence this may now occur within the brickwork. The crystallization of these salts can produce substantial expansive forces, causing spalling of the face of the brickwork. Also, under suitable conditions the moisture within the wall may evaporate from internal surfaces giving the impression that the wall continues to possess inadequate weather resistance.

Solid walls of clay brick can be subject to sulphate attack. This is more common in parapet and free-standing walls, including walls to chimney stacks. Bricks with a high soluble sulphate content are particularly vulnerable. General efflorescence can also be a problem, although this is rarely more than a deterioration in the appearance of the brickwork.

Older solid walls may have no DPC at ground level. Even those walls where this is provided may have incorrectly positioned DPCs; this can result in a range of defects. Parapet walls are particularly vulnerable. Where a DPC has not been provided immediately under a weatherproof coping, or has not been provided for the full width of the wall, then rain can penetrate the wall. The parapet wall may become saturated where the coping projection is inadequate or anti-capillary grooves or throatings have not been provided. It is suggested that a minimum projection of the coping of 40 mm is required with 25 mm clearance between the wall and the anti-capillary groove. The risk of chemical attack of the brickwork increases when it is saturated.

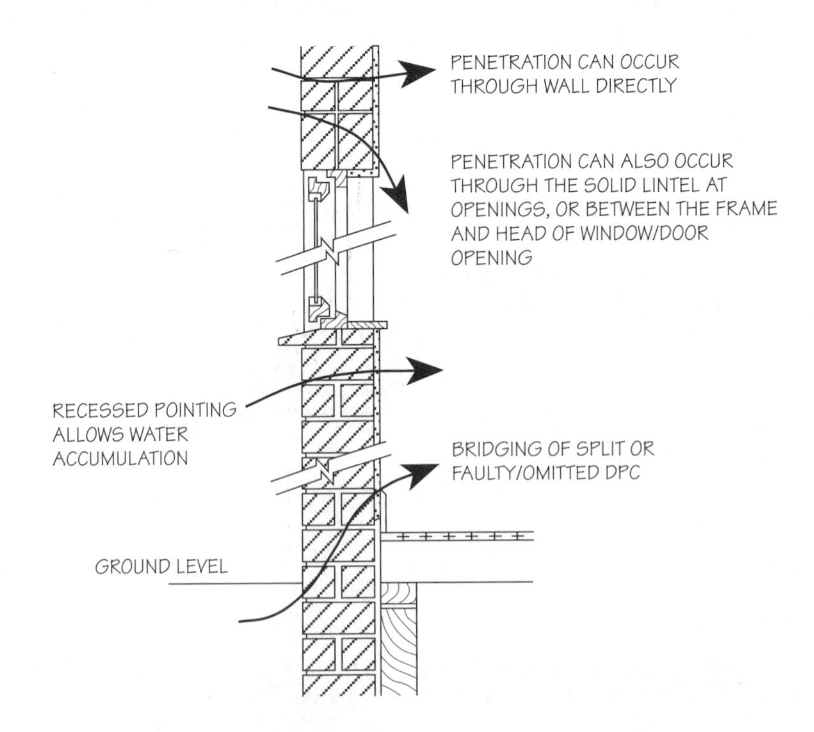


**Fig. 9.30.** Water penetration routes through solid masonry walling.

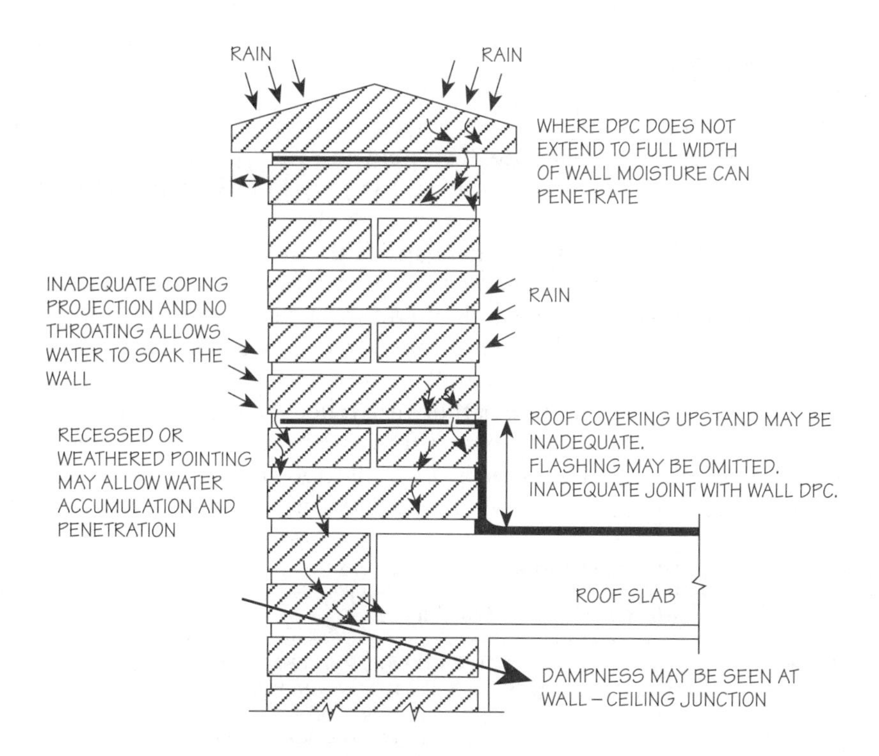


**Fig. 9.31.** Solid parapet wall: failure to resist dampness penetration.

Where a DPC is not provided at the base of the parapet wall there can be no continuity with the roof covering or flashing, and a pathway for rain penetration can exist. Even when a DPC is provided, where it does not pass through the entire thickness of the wall, rain may still enter the wall. This may produce damp patches on the ceilings near outside walls of upper rooms. For timber roofs then dampness in this region may also affect the durability of structural timbers around the parapet gutter.

**Fig. 9.32.** Severe diagonal cracking of a solid wall. The external supports appear to be ineffective, although they are evidence of a long-term problem.

## Solid walls – generally

The openings in solid walls to accommodate windows or doors may also be regions where rain may penetrate. The joint between the relatively impervious window or door frame and the wall is particularly vulnerable. Wind action and the lack of rain absorption over the door or window surface can cause rain to be blown through small gaps. This can increase the moisture content of timber frames and may cause deterioration of interior plasterwork around the opening. Rainwater run-off from the cill can cause similar problems to those associated with the parapet wall; in particular the projection of the cill and any anti-capillary throating can be critical.

In order to provide enhanced weather resistance solid brick walls may be rendered. Where this is applied over brickwork which has not effectively dried out, the entrapped moisture may disrupt the rendering. Render, although capable of significant durability and weather resistance, can fail

**Table 9.1** General defects in different types of wall

| Construction | Defect | Possible cause |
|---|---|---|
| Walls generally | Cracks | Tree damage, foundation movement, settlement, sulphate attack of mortar<br>High-strength mortars<br>Absence of movement joints |
| | Dampness | Penetration, rising |
| | Condensation | High thermal resistance and high permeability, no vapour barrier, nature of use of building, inadequate ventilation |
| | Efflorescence | Soluble salts in masonry |
| | Instability | Poor detailing and/or poor workmanship |
| Stone walls | Stone decay | Weathering and detailing |
| | Staining | Iron pyrites, corrosion of buried ties (also poor colour match) |
| | Delamination and spalling | Wrongly bedded cleavage planes, poor stone selection, differential movement between face and background<br>Loss of adhesion of stone veneer |
| | Dampness | No DPC or severe exposure<br>Poor detailing around openings. |
| | Sulphate attack | Atmospheric, or from internal rubble fill |
| Solid brick | Dampness | No DPC or severe exposure<br>Poor detailing around openings. |
| | Sulphate attack | Atmospheric, groundwater-borne, bricks (also causes bowing) |
| | Bowing | High slenderness ratio and absence or removal of lateral restraint |
| | Spalling | Frost action, expansion of free lime |
| Cavity wall | Dampness | Severe exposure, poor detailing<br>Cavity bridging and discontinuous DPCs, lack of weepholes |
| | Condensation | |
| | Interstitial | Lack of vapour barrier |
| | Surface | Impermeability, reduced ventilation |
| | Sulphate attack | Atmospheric, groundwater-borne, bricks |
| | Distortion | Corrosion of ties, lack of lateral restraint |

owing to a range of defects. Using tile hanging to achieve the required weather resistance may introduce a range of defects associated with vertical tiling.

Concrete lintels, beams and columns may pass through the thickness of the wall. The higher thermal conductivity of concrete may mean that these features form cold bridges having an increased risk of condensation.

## Revision notes

- Rain can pass through a solid wall; particularly in exposed locations. Open perpends, recessed pointing and missing or poorly positioned DPCs may offer routes for water penetration. Remedial treatments can trap water in walls. Junctions and structural openings provide other rain penetration pathways.
- Solid parapet walls can become defective where the DPC under the coping has been omitted or fails to cover the width of the wall. Short coping projections or a lack of anti-capillary weathering grooves can cause rain to saturate the wall.
- Solid walls of stone can fail owing to the range of defects associated with stone. Some stones are more porous than brick. Where stone is backed with brick, ferrous metal bonding ties and cramps may corrode.

## Discussion topics

- Compare the resistance to penetrating dampness of a solid masonry wall and a cavity masonry wall.
- Discuss the changes in the defects found in external masonry walls between 1900 and the present day.
- 'Since brickwork is basically porous there is little to be gained from providing DPCs in solid walls.' Discuss.
- Illustrate a possible mechanism of damp penetration through a solid parapet wall to a felt-covered flat roof.
- Compare the defects associated with stone solid walls with those of brickwork.
- Discuss the influence of openings in solid walls on the incidence of potential defects.

## Further reading

Addleson, L. (1992) *Building Failures – Guide to Diagnosis, Remedy and Prevention*, 3rd edn, Butterworth-Heinemann.

BRE (1964) *Design and Appearance – 1*, Digest 45, Building Research Establishment, HMSO.

BRE (1964) *Design and Appearance – 2*, Digest 46, Building Research Establishment, HMSO.

BRE (1975) *Decay and Conservation of Stone Masonry*, Digest 177, Building Research Establishment, HMSO.

BRE (1982) *External Walls: Reducing the Risk from Interstitial Condensation*, Good Building Guide 6, Building Research Establishment, HMSO.

BRE *Assessing Traditional Housing for Rehabilitation*, Digest 167, Building Research Establishment, ch. 7.

Cook, G.K. and Hinks, A.J. (1992) *Appraising Building Defects: Perspectives on Stability and Hygrothermal Performance*, Longman Scientific & Technical, London, ISBN 0 582 05108 8.

Mason, J. (1992) Solid masonry walls I – future technology. *Architects' Journal*, June.

Ransom, W.H. (1981) *Building Failures: Diagnosis and Avoidance*, 2nd edn, E. & F.N. Spon, London.

Richardson, B.A. (1991) *Defects and Deterioration in Buildings*, E. & F.N. Spon, London.

Taylor, G.D. (1991) *Construction Materials*, Longman Scientific & Technical, London.

# 9.10 General defects with cavity walls

## Learning objectives

You should understand the principal vulnerabilities of cavity masonry walls, affecting:

- structural stability and strength;
- resistance to damp penetration;
- overall durability.

## Cavity walls

Cavity walls were traditionally constructed with two brick leaves laid in stretcher bond. The rain resistance provided by the cavity was an improvement on the solid wall, but detailing became more difficult. Wide cavities were required to achieve a similar sound insulation to that of stone or solid brick walls, but this limited stability and thermal insulation. The cavity wall was frequently a compromise solution for several performance factors.

The principal issues related to deficiencies in cavity walls are damp penetration and rising damp; strength and stability generally, and around openings specifically; thermal performance, including occurrence of condensation; safety in fire where the cavity is not adequately closed; and overall durability.

## Damp penetration

Damp can penetrate the cavity wall via a faulty or omitted DPC, which should be used to protect the wall 150 mm above ground level and around openings to the wall. There may also be direct penetration of the wall fabric, or directed penetration of rainwater caused by obstacles in the cavity such as ties coated with trapped mortar or unattached (and therefore inclined) partial cavity insulation creating bridges. It is important to appreciate in considering such potential defects that the cavity zone can remain damp for large portions of the year, and that in driving rain conditions it has been demonstrated that rainwater can be driven into the cavity once the outer

*The Technology of Building Defects.* Dr John Hinks and Dr Geoff Cook.
Published in 1997 by E & FN Spon, 2–6 Boundary Row, London SE1 6HN, UK. ISBN 0 419 19770 2

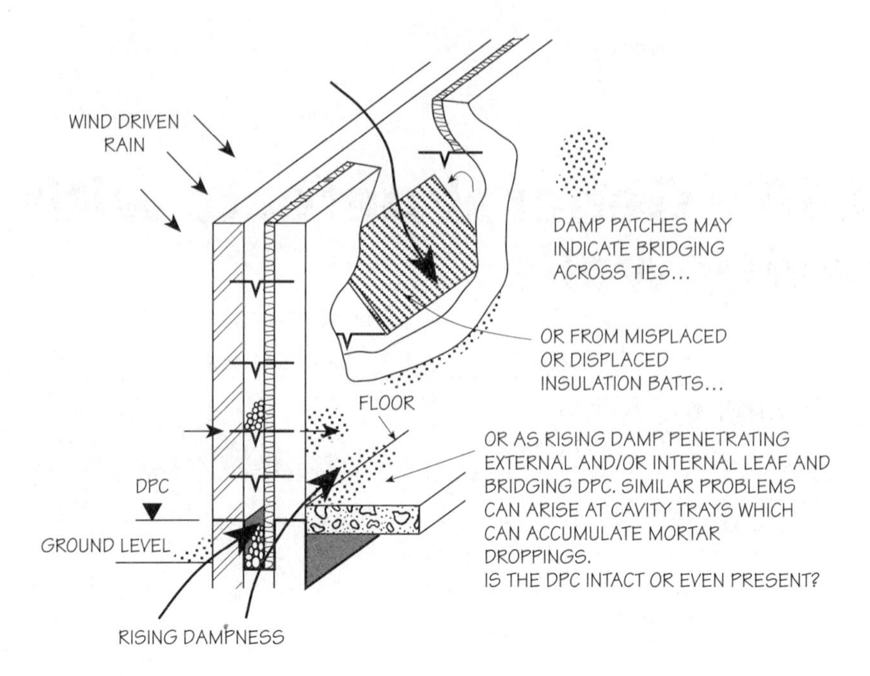


**Fig. 9.33.** Schematic representation of water penetration paths in cavity walls.

leaf is saturated. Clearly, in such instances the thermal performance of the wall will not be as expected.

The risk of rain penetration is related to a number of factors, the importance of which depends also on the exposure of the wall. Consider the following factors when assessing the causal path/factors of a water penetration fault.

- The nature and disposition of insulation in the cavity, which varies in material composition and resistance to damp penetration and also in the extent to which it fills the cavity zone.
- Consider also whether insulation may be bridging the DPC within the cavity, and thereby providing a path for moisture transport.
- As to be expected, the cavity width has a bearing on the scope for water bridging, but it is also important, for any cavity, to consider the possibility of inclined ties or mortar trapped on ties, which provide either a direct or capillary path for moisture ingress. Ties allowing water passage may produce localized patches of damp at any position on the inner wall. Another mechanism can occur where an insulation batt is detached from the tie clip and forms an inclined path for water penetration. In such cases there will also be focused patches of damp. In both instances the appearance and severity of the damp penetration may be related to the pattern of rainfall.
- Mortar droppings sitting in the base of the cavity can also provide a route for more constant damp penetration, perhaps worsening after heavy rainfall.
- Other problems with water penetration could lie in the detailing of components and/or openings in the walls, and of connections between elements. Hence poor overhangs and/or weathering of parapet copings, soffits, window cills and external door steps may provide routes for

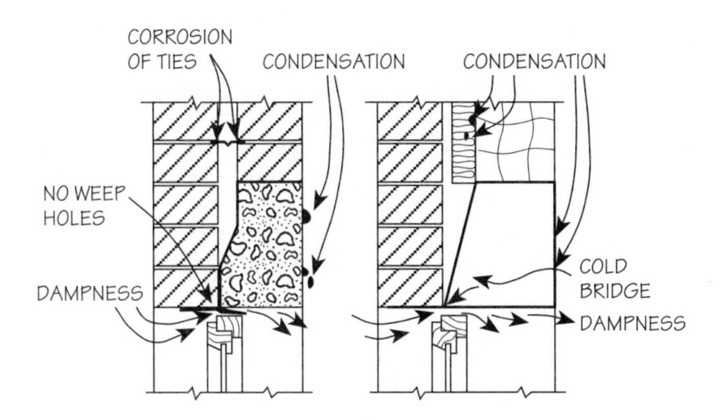


**Fig. 9.34.** Possible water penetration routes at openings in cavity walls.

water penetration. Windows and doors may provide poor direct weather-proofing, but can also contribute to water penetration at their perimeter connections, especially if such joints are not adequately sealed. Check such dampness carefully to ascertain whether it is water penetration or condensation occuring at a cold bridge. The pattern of occurrence may assist in such a diagnosis.

## Strength and stability generally

Aside from the major movement of a building which usually manifests itself as cracking, masonry cavity walls have a series of potential deficiencies associated with their general strength and stability.

The significance of slenderness and connectivity between the two leaves of the cavity wall is critical. Untied leaves will act independently and the inner structural leaf will be denied the bracing usually afforded by the other leaf being connected to it. The stability of walls in general is dependent upon their height-to-thickness ratio – the slenderness ratio – and a structural discontinuity across the cavity can result in both leaves being excessively slender in terms of stability. In addition to the static implications, the capacity for dynamic loading of the walls will be impaired – the (untied) outer leaf will carry wind loads alone. Hence cavity ties are critical to the stability of the wall, and are covered elsewhere as a stand-alone item in this book.

Walls are buttressed by the floors and roof to which they are connected, turning the structure of the building into a 3D box rather than a collection of 2D elements. Hence the connectivity between the floors (and roof) and the walls is critical also, and strapping should be checked for a number of potential deficiencies which can render the restraint minimal (Figure 9.35). Such defects can be the result of poor design/specification or workmanship. Note that this means conversely that overloading or failures in the floors or roof will result in extra imposed loads on the walls, which may be eccentric or unbalanced, thereby causing potentially serious secondary defects in the wall.

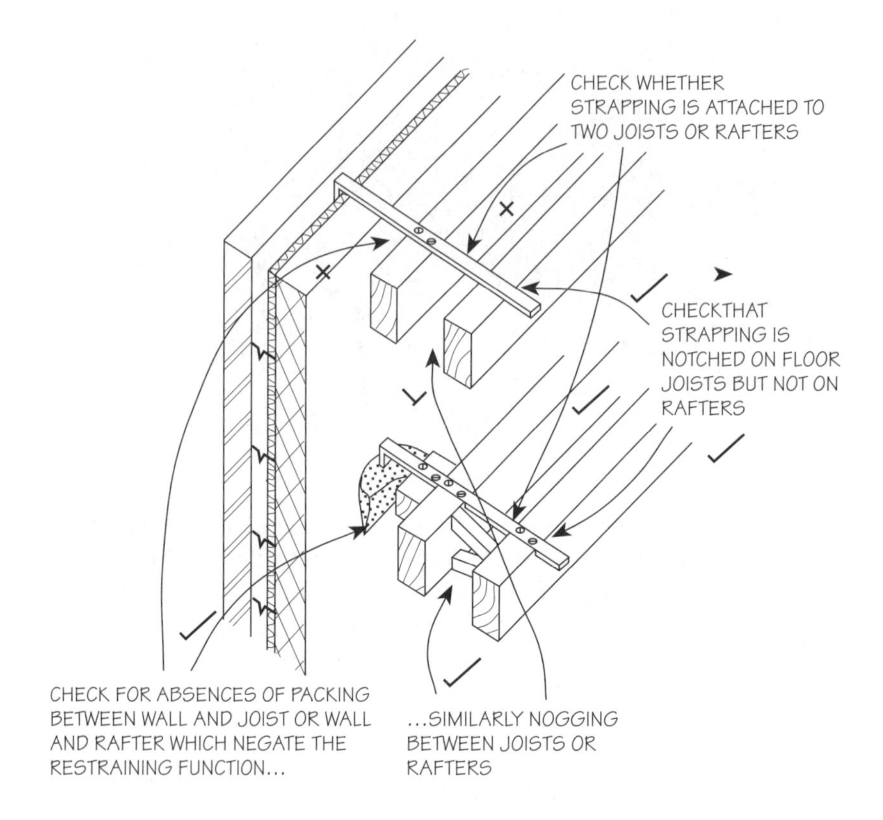


**Fig. 9.35.** Schematic contrast of sound and unsound lateral restraint provision to walls by floors and roof.

It is common to find minor cracking in modern rigid cavity walling, and in the case of calcium silicate brickwork it is quite probable. Compared with the older, more ductile masonry construction, modern domestic cavity wall construction is intolerant of movement stresses and tends to exhibit cracking more readily as a means of relieving the stresses. The distinction between minor and major cracking is a matter of local judgement – the patterns of cracks may help the assessor appraise the cause and effect with confidence, and it is essential to consider all cracking seriously.

Masonry walls will exhibit reversible thermal and moisture movement, which, if there is no accommodation for this movement in the form of movement joints, may relieve the stresses by displacement and/or cracking in the wall (see elsewhere). The width of the joints and spacing are important to allow the cumulative expansion and contraction induced by moisture and thermal movement to be absorbed without overstressing the sealant. There is also the possibility of irreversible movement in bricks fresh from the kiln as they absorb moisture. Careful site planning can avoid the problems that this can cause in a new building.

Specific structural problems can appear at or around openings, which are a point of weakness in walls and a common focus for cracking relief of stresses. The transfer of load around openings can cause localized problems, especially with narrow piers between window openings. These parts of walls can become overstressed, and may not provide adequate bearing for both lintels. In some instances, especially with upper windows with their head lying at the eaves, the lintel may be omitted entirely, thereby transferring the load to the wall plate. Inspection is advisable here.

## General durability issues

The use of exposed concrete lintels and cill was a common practice some years ago, and this creates ongoing problems with cold bridging where the lintel or cill crosses both walls. It can also lead to direct water penetration problems. Externally, the durability of the exposed concrete can be problematic, with deterioration as a result of corroding reinforcement producing spalling, unsightliness and (more significantly) structural weakness. Partial failure may be obvious from bulging prior to spalling and by tapping the surface to see if it is sound or hollow. Wall tie durability can be a problem also. Old or ungalvanized ties may cause disruption to the wall as they corrode (especially the thicker fishtail type); however, this is not necessarily a symptom. Internal inspection may be needed to identify even advanced failure in thin ties, which may present no external symptoms prior to failure.

Surface rendering can be a useful indicator of faults in the wall, especially cracking which may be mirrored in the rendering. Rendering also will have its own durability problems, associated in part with exposure.

Brickwork damage can occur (owing to deterioration of the bricks, especially below DPC level because of frost attack. Sulphate attack can affect brickwork and/or mortar (see elsewhere).

## Revision notes

### Strength and stability

- Slenderness problems will occur with walls which have an inappropriate height/thickness ratio, creating instability in the wall. These can arise when cavity wall ties fail, leaving the two leaves unconnected.
- Connections with floors and roofs are critical to provide lateral restraint to the wall. A number of installation defects can arise with such connections, including the omission of packing between floor joists (or roof timbers) running parallel to the wall and the wall which they should be stabilizing, the omission of packing between the outermost pair of joists (or rafters) used for lateral restraint and the absence or improper attachment and notching of the strapping to the joists (or rafters) and/or wall.
- Minor cracking is not uncommon in modern cavity walls, and may be due to material behaviour (especially with calcium silicate brickwork). It is important to distinguish between minor and major cracking by careful analysis.

### Damp penetration

- Rain can penetrate the cavity, and in the UK climate the cavity zone may remain damp for long periods of time.
- There is a range of potential vulnerabilities to rain penetration associated with the type of cavity insulation, structural cavity width, the mortar mix joint finish and pointing profile, and any applied external finishes, all of which need to be considered in conjunction with the exposure of the wall.
- Problems may appear as damp patches on the internal wall, the patterns of occurrence being related to the location of DPCs, wall ties and/or cavity trays. Consider also the timing of appearance in relation to rainfall.
- Cavity trays may cause problems with overspill or leaks at damaged points. Mortar droppings can be a causal component.

### Durability generally

- Check lintel presence and condition. Steel lintels can corrode, concrete lintels can carbonate and/or suffer corroded steel reinforcement.
- Wall tie failure may produce cracking symptoms (see separate section), but not necessarily.
- Rendering defects may be a tell-tale for wall defects.
- Brickwork deterioration occurs with frost, sulphate attack to the bricks and/or mortar.

## Discussion topics

- How can damp penetration be distinguished from condensation at a cold bridge? Discuss the mechanisms that may be involved and produce a set of guidelines for distinguishing between the two sources of dampness in buildings.
- What faults may appear as a result of inadequate lateral restraint to walls?
- Produce an analysis of the factors affecting the resistance to damp penetration of a modern cavity wall and rank them in order of importance.
- Describe the problems with damp penetration at cavity trays and how this may be identified in practice.
- Discuss how the thermal insulation requirements of cavity walls may be compromised by other defects.

## Further reading

Addleson, L. (1992) *Building Failures: Guide to Diagnosis, Remedy and Prevention*, 3rd edn, Butterworth-Heinemann.
BDA *Cavity-Insulated Walls: Specifiers Guide*, Brick Development Association Good Practice Note 2.
BRE (1983) *External Masonry Walls: Vertical Joints for Thermal and Moisture Movement*, Defect Action Sheet DAS 18, Building Research Establishment.
BRE (1983) *External and Separating Walls – Lateral Restraint at Intermediate Timber Floors: Specification*, Defect Action Sheet DAS 25, Building Research Establishment.
BRE (1983) *External and Separating Walls – Lateral Restraint at Intermediate Timber Floors: Installation*, Defect Action Sheet DAS 26, Building Research Establishment.
BRE (1983) *External and Separating Walls – Lateral Restraint at Pitched Roof Level: Specification*, Defect Action Sheet DAS 27, Building Research Establishment.
BRE (1983) *External and Separating Walls – Lateral Restraint at Pitched Roof Level: Installation*, Defect Action Sheet DAS 28, Building Research Establishment.
BRE (1990) *Assessing Traditional Housing for Rehabilitation*, Report 167, Building Research Establishment, ch. 8.
Duell, J. and Lawson, F. (1983) *Damp Proof Course Detailing*, Architectural Press.
Endean, K. (1995) *Investigating Rainwater Penetration of Modern Buildings*.
Newman, A. (1933) *Rain Penetration Through Masonary Walls – Diagnosis and Remedial Measures*, Report 117, Building Research Establishment.
NHBC (1995) *External Masonry Walls*, Standard, National House Building Council, ch. 6.1.

# 9.11 Cavity wall tie problems

## Learning objectives

You should understand:

- the mechanism of failure and the common symptoms of wall tie corrosion;
- the role of wall ties in specific locations such as around openings;
- the significance of partial or total failure for the building.

The generally accepted period for the introduction of wall ties is around 1923, although earlier properties may have wall ties and cavity walls, and the uptake was somewhat regionally oriented. It is possible to find large numbers of houses built in London in the 1930s with solid walls for instance.

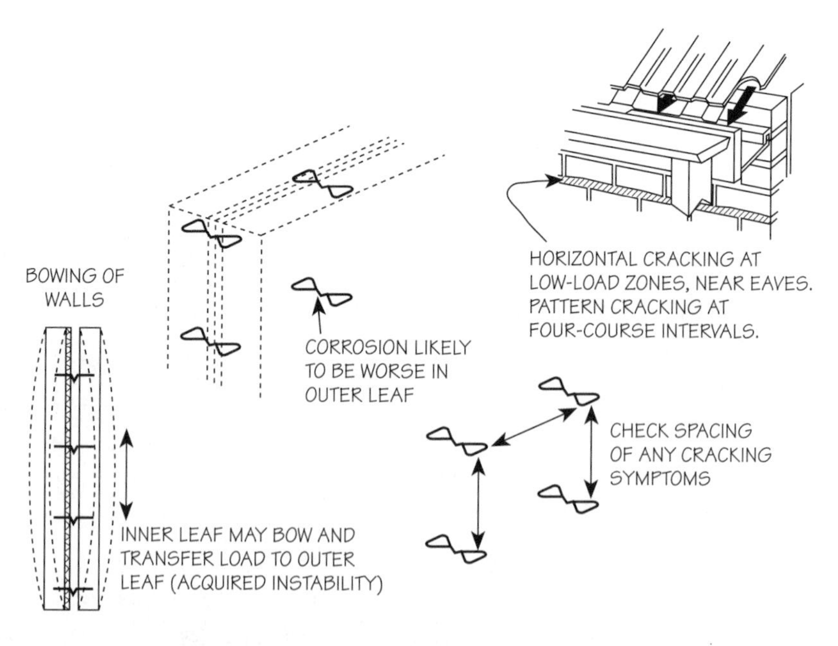


**Fig. 9.36.** Symptoms of wall tie corrosion.

*The Technology of Building Defects.* Dr John Hinks and Dr Geoff Cook.
Published in 1997 by E & FN Spon, 2–6 Boundary Row, London SE1 6HN, UK. ISBN 0 419 19770 2

Note that some early cavity construction was disguised as solid Flemish bond by using snap header bricks, which may mean that the loadbearing capacity of the wall and stability are less than they would appear.

The general large-scale failure of cavity wall ties has not yet occurred, although serious concerns have been expressed. The early galvanized steel ties were prone to rusting, and revisions to the treatment requirements for steel ties (in 1982) have resulted in thicker galvanized coatings. Prior to galvanizing, the technique was bitumen-casting.

The life expectancy of galvanized wall ties is directly related to the coating thickness. Damage or patchy coating can lead to premature failure, and there have been indications that the life expectancy of the early ties can be less than that assumed for the cavity wall.

Cavity walls in which the mortar suffers from carbonation will have a higher risk of corrosion in the wall ties, since the carbonation reduces the protective alkalinity of the mortar surrounding the tie.

**Fig. 9.37.** Wind damage to a cavity wall forming the gable end of a domestic building. Significant negative pressure on the face of the wall has detached it from the inner leaf, overcoming any resistance of the wall ties. (Reproduced from the *Kentish Gazette*.)

Rusting failure of the sections of the ties within the leaf produces expansion between the courses, usually more so in the external (damp) leaf. Highly exposed walls, and early wall construction using sulphate-high black ash mortar or more modern chloride-based mortar are most susceptible. The porosity of the mortar will also be critical since frequent wetting of the ties accelerates their corrosion. Hence cracked rendering, which may keep the wall wet over prolonged periods may accelerate failure. There has been debate over whether full-fill non-absorbent cavity insulation contributes to failure, and under what, if any, conditions it may be identified as a critical factor.

The effect on the wall of wall tie failure will be horizontal pattern cracking corresponding to the location of the wall ties (usually every four courses and approximately 5 mm wide). There may also be vertical and/or diagonal cracking. The horizontal cracking is most likely to occur where the resistance against uplift is least, such as near eaves.

However, it is possible for the tie to corrode within the cavity zone. This is less symptomatically to the building structure than corrosion within the outer leaf, but gives no warning symptoms.

The significance of wall tie failure will depend on the form and shape of the walls and their inherent stiffness. Large slender walls, such as gables, will be most vulnerable. Walls which are in stable equilibrium may not be visibly affected by wall tie failure; however, the two leaves will act independently rather than as a tied unit. In high wind conditions they may be particularly vulnerable to failure. Since gables may be subject to high exposure also, they are particularly prone, in terms of exposure and the wind forces which may cause their collapse in instances of wall tie failure. Other vulnerable areas include any supported brickwork (such as that between windows). Clearly, walls that have already distorted may fail when the ties fail.

## Revision notes

- May occur within the cavity zone or within the wall (usually affects the outer leaf).
- Disruption is the result of expansion of wall tie as it corrodes.
- Symptoms may include vertical and/or diagonal cracking, but are usually horizontal cracking, approximately every four courses. Cracking occurs most in lightly loaded areas such as eaves and gables. In cases where the corrosion occurs within the cavity there may be no external symptoms.

## Discussion topic

- Identify the symptoms of wall tie failure and describe the process you would use to distinguish wall tie failure from other causes of similar symptoms.

## Further reading

Addleson, L. (1982) *Building Failures – A Guide to Diagnosis, Remedy, and Prevention*, Architectural Press.

Atkinson, G. (1985) New materials – innovations are not a main cause of building failure. *Structural Survey* **6**(2), 113–17.

BRE (1983) *External Masonry Cavity Walls: Wall Tie Replacement*, Defect Action Sheet DAS 21, Building Research Establishment.

BRE (1988) *External Masonry Cavity Walls: Wall Tie* – Selection and Specification, Defect Action Sheet DAS 115, Building Research Establishment.

BRE (1988) *External Masonry Cavity Walls: Wall Tie* – Installation, Defect Action Sheet DAS 116, Building Research Establishment.

BRE (1990) *Assessing Traditional Housing for Rehabilitation*, Report 167, Building Research Establishment, ch. 8.

BRE (1993) *Installing Wall Ties in Existing Construction*, Digest 329, Building Research Establishment.

BRE (1995) *Replacing Wall Ties*, Digest 401, Building Research Establishment.

de Vekey, R.C. (1979) *Corrosion of Steel Wall Ties: Recognition, Assessment, and Appropriate Action*, BRE Information Paper, IP 28/79, Building Research Establishment, HMSO.

de Vekey, R.C. (1990) *Corrosion of Steel Wall Ties: History of Occurrence, Background, and Treatment*, Building Research Establishment.

de Vekey, R.C. (1990) *Corrosion of Steel Wall Ties: Recognition and Inspection*, Building Research Establishment.

Eldridge, H.J. (1976) *Common Defects in Buildings*, HMSO.

Hollis, M. (1990) *Cavity Wall Tie Failure*, Estates Gazette Professional Guides.

Moore, J.F.A. (1981) *Performance of Cavity Wall Ties*, Information Paper IP 4/81, Building Research Establishment.

Seeley, I.H. (1987) *Building Maintenance*, 2nd edn, Macmillan Building and Surveying.

Williams, P.H. (1986) *Cavity Wall Tie Corrosion*, Surveyors Publications.

# 9.12 Problems with cavity trays and weepholes

## Learning objectives

- You should be able to appreciate that cavity trays, whilst installed to resist the passage of water into the building, may in practice achieve the opposite effect.
- You will identify some of the implications of their omission from cavity walls and inadequate jointing with other DPC's or flashings.
- You will be made aware of some of the defects associated with weepholes.

## General problems

Cavity trays are used to stop penetrating dampness from entering the inner skin of cavity walls and to divert the penetrating dampness towards the outer skin of the cavity wall. The greater the permeability of the outer leaf the greater the amount of rain penetration. The cavity tray can also offer protection from penetrating dampness above window and door openings and other areas where, because of constructional details, the cavities are closed.

Dampness penetration may occur where cavity trays are not provided in the following locations:

- where cavity walls are built directly off concrete floors or beams;
- where external leaves of cavity walls become internal walls, at a lower level;
- above lintels, air bricks and meter boxes or other areas where cavities are bridged;
- at the base of cavity parapet walls.

Other particular problems include the following.

- Insufficient upstand to cavity tray, typically $> 75$ mm upstand may be bridged by mortar droppings. These may close the cavity at the cavity tray.

*The Technology of Building Defects.* Dr John Hinks and Dr Geoff Cook.
Published in 1997 by E & FN Spon, 2–6 Boundary Row, London SE1 6HN, UK. ISBN 0 419 19770 2

- Any damage to the cavity tray material which could have been caused during the construction of the cavity tray.
- No mortar haunching to support the cavity tray. The traditional cavity tray was metal, zinc, copper or lead, and haunched. New, flexible materials are unhaunched and can therefore sag and distort.
- Inadequate lapping, sealing or jointing along the length of the cavity tray. Simple overlapping is unlikely to be effective.
- Stop ends may be either poorly formed or missing.

## Penetration of the cavity tray

Penetration can occur under the cavity tray where it is inadequately bedded. Problems can occur when the cavity tray goes under any flashings or DPC's associated with the detail. Failure to extend the cavity tray in front of the window head may divert water down the face of the frame or onto the back of the frame, where it may accumulate in weathering grooves.

Inadequate overlap of the cavity tray and the vertical DPC to the jamb of the opening may allow rain to penetrate.

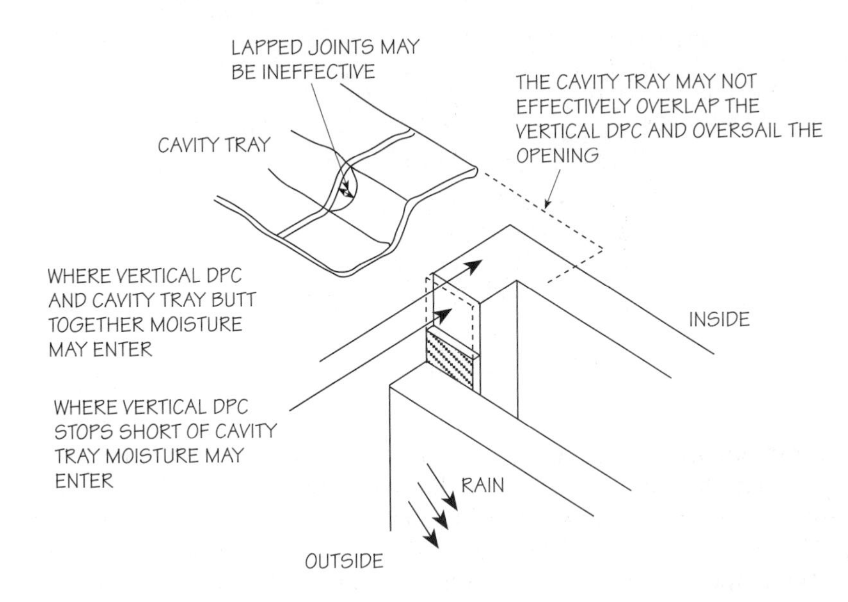


**Fig. 9.38.** General defects – cavity tray and vertical DPC junctions.

Where cavity trays meet columns rain may enter where the stop ends are not sealed to the columns.

The common use of steel lintels, over door and window openings, which also perform the function of a cavity tray, has the potential to solve some of the problems with flexible cavity trays. Where they fail to provide corrosion resistance their long-term strength may reduce. On-site cutting of standard-length lintels may result in poorly protected surface edges. Corrosion in these areas may penetrate under galvanized or other protective coating owing to differential oxygenation.

Steel lintels may also produce cold bridges and cause surface condensation

## Parapets

The cavity tray at the base of the parapet wall may fail to stop dampness where it does not extend through the full thickness of the external skin of the cavity wall. The cavity tray may not overlap the flashing to the roof covering. Flashings may not penetrate the full thickness of the inner leaf and therefore offer little resistance to damp penetration. Where flashings are chased into the same bed joint as the cavity tray, any joint between the two may fail to resist penetrating dampness.

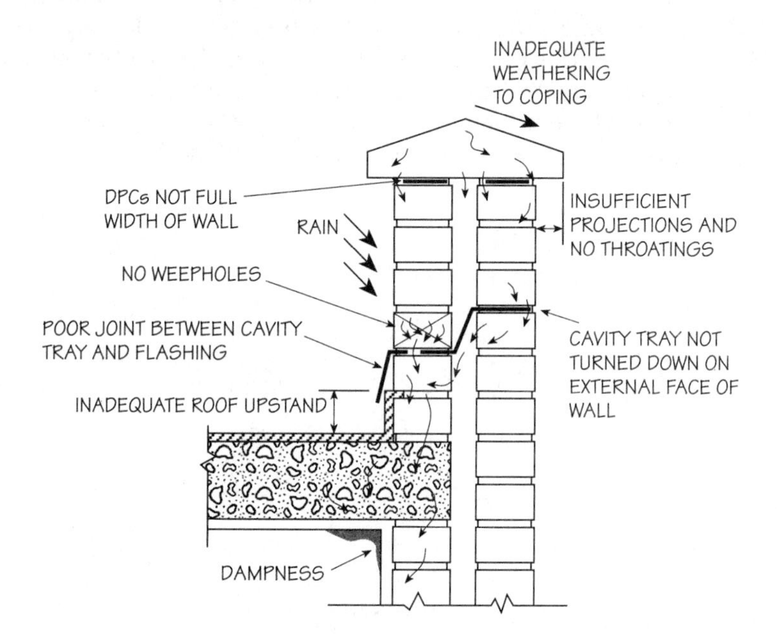


**Fig. 9.39.** Parapet wall – cavity tray problems.

The common method of sloping the cavity tray towards the inner leaf may provide a path for damp penetration. If the cavity tray in the outer leaf does not extend across the full width of the wall and is turned down to form a projecting drip, rain may drive under the cavity tray in the wall and run down the slope in the cavity to enter the inner leaf. This can cause damp patches to appear at the perimeter of upstairs rooms and may increase the moisture content of roof timbers in the eaves region.

## Weepholes

Essential to provide and keep open, since they can get blocked up. Where blockage occurs then water can back-up and move through the poor joints in the cavity tray to the inner leaf or fill the cavity up to the next mortar joint where it may escape. If this is above the height of the cavity tray then water may come in contact with the inner leaf.

Care is needed since in exposed locations rainwater can be blown into the cavity through weepholes.

Drilling weepholes in existing walls can damage the cavity tray and may cause debris to accumulate in the cavity

Providing weepholes at low level can help in damp-proofing of cavity walls. They can reduce vapour pressures within the cavity. Unfortunately they can provide a pathway for water splashed up from outside the wall to enter the cavity. High ground levels may bury weepholes.

## Insulated cavities

Cavity trays are needed where insulation on the inner skin of cavity walls, or insulation batts which fill the thickness of the cavity, do not extend to the full height of the wall. Rain can penetrate the outer skin, run along the wall tie or mortar-coated cavity tie and drip into the exposed top edge of the insulation board or batt.

Where insulation boards are inadequately fixed to the outside surfaces of the inner leaf of cavity walls, they may fall and bridge the cavity. Mortar extrusions from the outer leaf may partially bridge the cavity, allowing water to drip onto poorly fixed or distorted insulation boards. Where the cavity is filled with insulation batts the mortar extrusions can compress the insulation material and any water entering the batts may follow the compression layers towards the inner leaf of the wall.

Insulated cavities may block or close weepholes, rendering them partially or completely ineffective. Where insulation prevents water from draining freely from the cavity this will influence the effectiveness of the weepholes.

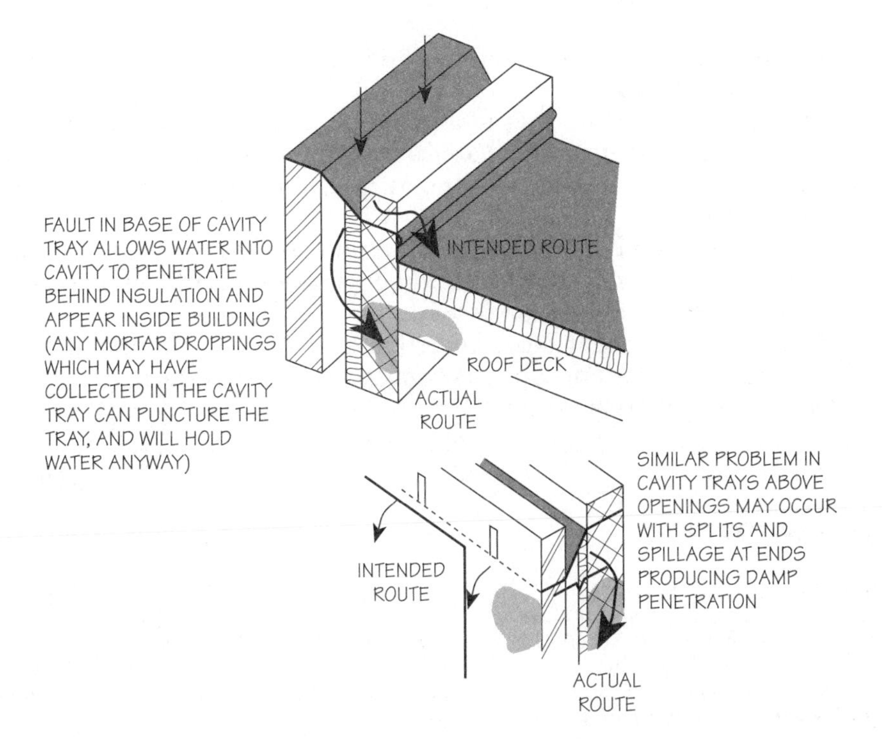


**Fig. 9.40.** Possible defect path for water penetration to insulation: cavity tray failure.

Retro fitting cavity insulation may cause the cavity below the level of the DPC to be filled. This may allow ground water to move up through the wall.

## Revision notes

- Cavity trays may allow dampness to penetrate cavity walls owing to their omission from areas adjacent to bridges across the cavity or above cavity insulation material. An insufficient upstand of the cavity tray where mortar droppings may accumulate to close the cavity can also fail. The cavity tray should be an integral part of DPC provision around an opening.
- The flexible unhaunched cavity tray materials can sag and distort allowing water to pass through the open joints. A cavity tray which goes under any flashings or DPC's can direct water into the wall. There may be an inadequate overlap between the cavity tray and vertical DPCs around openings.
- Rain can be blown through weepholes and fully insulated cavities can close weepholes. Blocked weepholes allow water to back-up and pass through poor cavity tray joints or fill the cavity with water.

## ■ Discussion topics

- Compare the nature and extent of defects associated with rigid and flexible cavity trays.
- Ilustrate three failure mechanisms of cavity trays in an uninsulated wall and comment on the water penetration pathways.
- Discuss the influence of weepholes on the defects associated with cavity walls.
- There are damp patches appearing on the internal finishes at either end of a window opening. Identify possible causes of the dampness and describe an inspection procedure.
- 'The relatively recent move to produce solid walls, by filling cavity walls with insulation, suggests that the range of defects associated with solid walls will return.' Discuss.

## Further reading

BRE (1982), *Cavity Trays in External Cavity Walls: Preventing Rain Penetration*, Defect Action Sheet 12, Building Research Establishment, HMSO.

BRE (1987) *Cavity Parapets – Avoiding Rain Penetration (Design)*, Defect Action Sheet 106, Building Research Establishment, HMSO.

BRE (1987) *Cavity Parapets – Installation of Copings, DPC's, Trays and Flashings (Site)*, Defect Action Sheet 107, Building Research Establishment, HMSO.

Cook, G.K., and Hinks, A.J. (1992) *Appraising Building Defects, Perspectives on Stability and Hygrothermal Performance*, Longman Scientific & Technical, London.
Duell, J. and Lawson, F. (1983) *Damp Proof Course Detailing*, Architectural Press, London.
Poutney, M.T., Maxwell, R. and Butler, A.J. (1988) *Rain Penetration of Cavity Walls: Report of a Survey of Properties in England and Wales*, Information Paper 2/88, Building Research Establishment, HMSO.
PSA (1989) *Defects in Buildings*, HMSO.
Ransom, W.H. (1981) *Building Failures Diagnosis and Avoidance*, 2nd edn, E. & F.N. Spon, London.
Richardson, B.A. (1991) *Defects and Deterioration in Buildings*, E. & F.N. Spon, London.

CHAPTER 10

# Movement and distortion problems in walls generally

A series of reviews of generic defects with walling.

## Learning objectives

- You should become familiar with the range of potential defects in walls, which are commonly the cause of cracking.
- An understanding of the nature of the forces involved and the types of cracking symptom will be needed to distinguish correctly between the various movements and distortion problems in walls.

**Fig. 10.1.** Diagonal cracking and displacement of a brick retaining wall due to differential movement between the base and the top of the wall.

**Fig. 10.2.** Movement of the structure which can occur at DPC level may also be visible internally.

# 10.1 Ballooning in walls

## Learning objectives

- One of the symptoms of unbalanced forces in walls, this defect is important to the stability of domestic-scale construction, and requires a relatively ductile construction type if it is to be tolerated.
- It is important to understand the relationship between the forces in the building structure.
- Ballooning is a relatively common defect with which you should be familiar.

Ballooning in walls is a symptomatic response to restrained expansive or outward forces. There are two principal causal modes of ballooning – expansion within the wall and acquired lateral instability.

In the former instance, the expansion may be caused by sulphate attack. This is usually predominant in the outer leaf or zone of a brick wall, where there is the greatest supply of moisture to support ongoing sulphate attack. The expansion of the outer leaf or zone produces a bulging in the wall.

**Fig. 10.3.** Ballooning of the gable end of a stone building. The downpipe acts as a visual reference line. (University of Reading.)

*The Technology of Building Defects.* Dr John Hinks and Dr Geoff Cook.
Published in 1997 by E & FN Spon, 2–6 Boundary Row, London SE1 6HN, UK. ISBN 0 419 19770 2

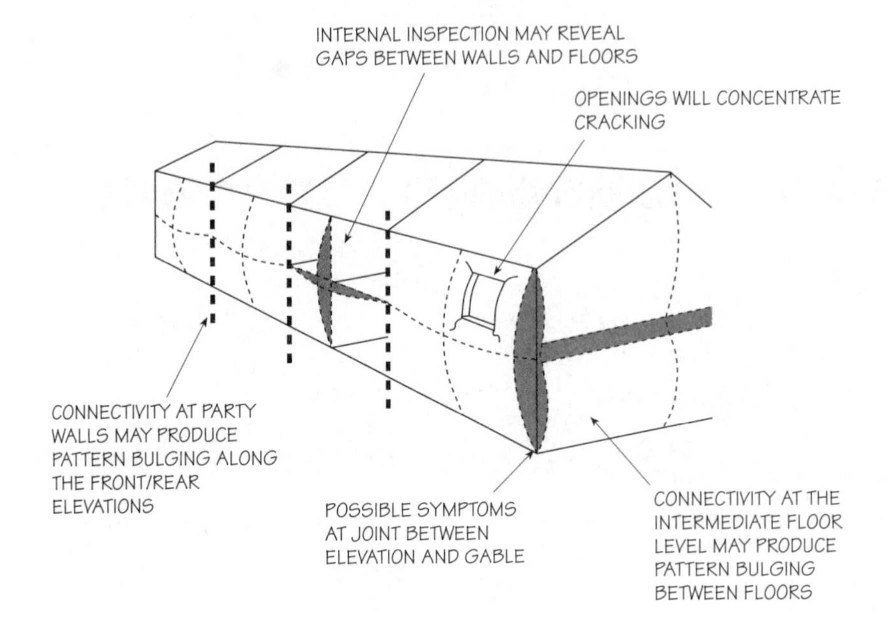


**Fig. 10.4.** Ballooning in terraced housing.

The other principal cause is where the resistance to vertical movement is greater than the resistance to lateral movement (connectivity). Loads such as a sagging roof or overloaded floors can lead to ballooning in the elevation or gable.

Such deformations can lead to problems with support for intermediate floor joists and internal walls being reduced. There will also be the risk of roof and wind loads not being within the capacity of the deformed wall.

The extent of deformation will be affected by the connectivity of the wall with the remainder of the building. Where the connectivity is high the ballooning may emerge as pattern bulging between zones of connectivity. In terraced housing this may occur between the party wall connections.

## Revision notes

- A symptom of outward or restrained vertical forces in the external wall.
- The result of sulphate attack within the external wall causing differential expansion across its thickness, or
- the reaction of the wall to excessive forces vertically or horizontally.
- Ballooning may occur in the horizontal or vertical plane.

## ■ Discussion topics

- Produce an annotated section through the front elevation of a house suffering from ballooning, identifying the likely symptoms and producing a checklist for appraisal.
- How would you distinguish between ballooning induced by sulphate attack and by a reaction to unbalanced forces?

## Further reading

Richardson, C. (1985) Structural surveys: 2. Data sheets, general problems. *Architect's Journal* **182**(27), 63–71.
Richardson, C. (1985) Structural surveys: 4. Common problems 1850–1939, *Architect's Journal*, **182**(29), 63–69.

# 10.2 Bulging in walls due to inadequate restraint

Bulging or ballooning in walls can be caused by overloading of floors which transmit these loads under deflection as inclined forces on the walls. Bulging may also be exacerbated by a lack of restraint from the floor structure if the appropriate corrections to parallel joists are missing. The effect in both cases may be a full-height bulging of the wall – the cause of which will require investigation for positive identification. Note that a wall which is bulging may pull the parallel connected floor joists into deflection also.

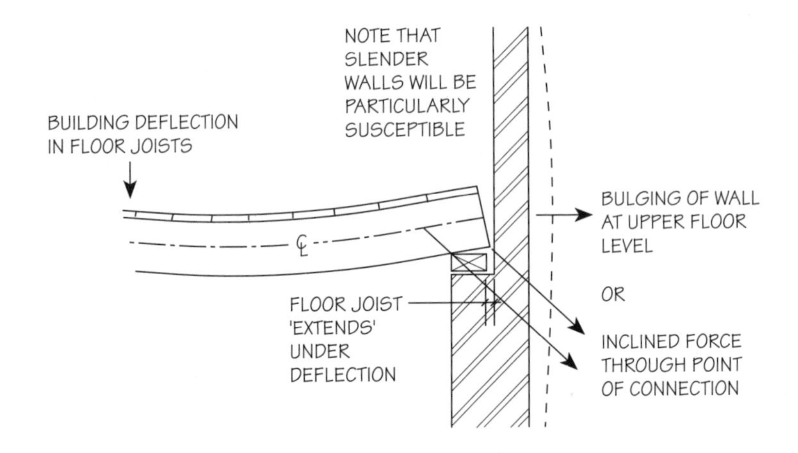


**Fig. 10.5.** Side forces from bending timber joist or beam causing wall to bulge.

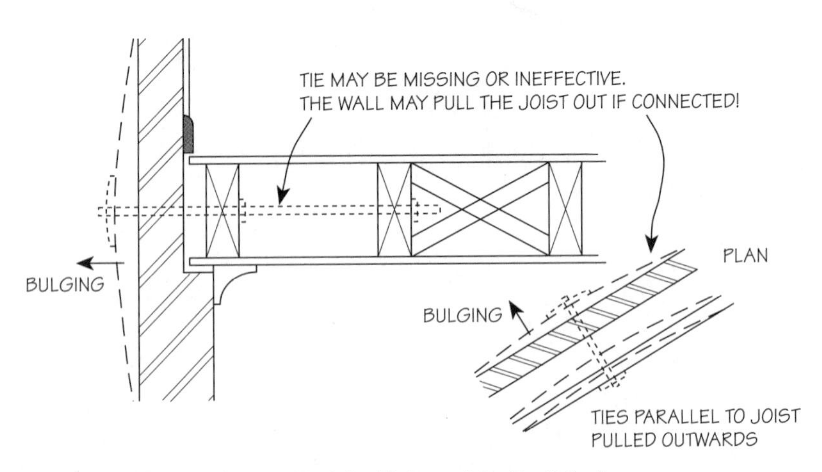


**Fig. 10.6.** Lack of support from parallel joists allows bulging of wall.

*The Technology of Building Defects.* Dr John Hinks and Dr Geoff Cook.
Published in 1997 by E & FN Spon, 2–6 Boundary Row, London SE1 6HN, UK. ISBN 0 419 19770 2

# 10.3 Tensile cracking in brick walls

The tensile capacity of brickwork is usually relatively low. It is also normal for the mortar to have a lower tensile capacity than the brickwork, and therefore to exhibit the cracking. In cases of brickwork with very strong mortar the cracking may pass through the bricks.

The problem of cracking in brickwork is also determined by the position of openings and other weaknesses in the elevation. Wherever the tensile forces concentrate, the cracking will occur. This will be at the point of least resistance, and will therefore tend to pass through openings.

Tensile cracking can be the result of drying shrinkage (reversible or irreversible), which can affect concrete or calcium silicate bricks as they dry out after manufacture (an irreversible movement) and all bricks as they pass through thermal movement cycles.

The tensile crack does not usually cross the DPC, since brickwork below the DPC tends to keep a fairly constant moisture content. The DPC also tends to act as a slip membrane.

In cases of reversible movement, perhaps caused by seasonal shrinkage/ swelling cycles in the structure or substructure, the cracks will change in size. They may close up at certain times of the year according to their cause. Cracks produced by irreversible dimensional changes alone in materials will not change.

A specific problem arises with calcium silicate brickwork.

*The Technology of Building Defects.* Dr John Hinks and Dr Geoff Cook.
Published in 1997 by E & FN Spon, 2–6 Boundary Row, London SE1 6HN, UK. ISBN 0 419 19770 2

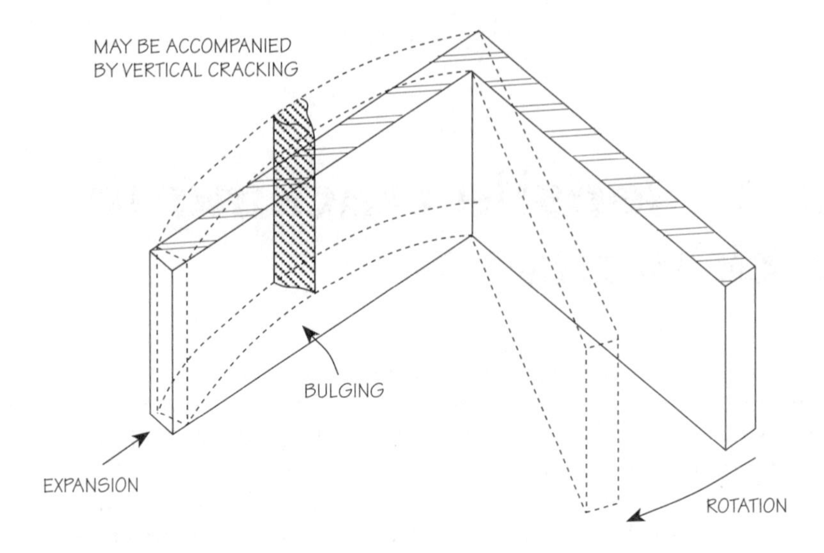


**Fig. 10.7.** Rotation and cracking in walls: bulging of wall attempting expansion against restraint. This phenomenon may occur with ballooning in elevations and gables.

**Fig. 10.8.** Tensile cracking – even through bricks where a high mortar strength is present. (K. Bright.)

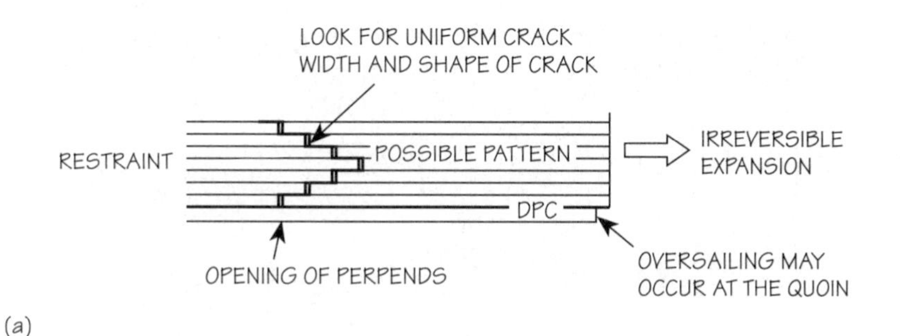


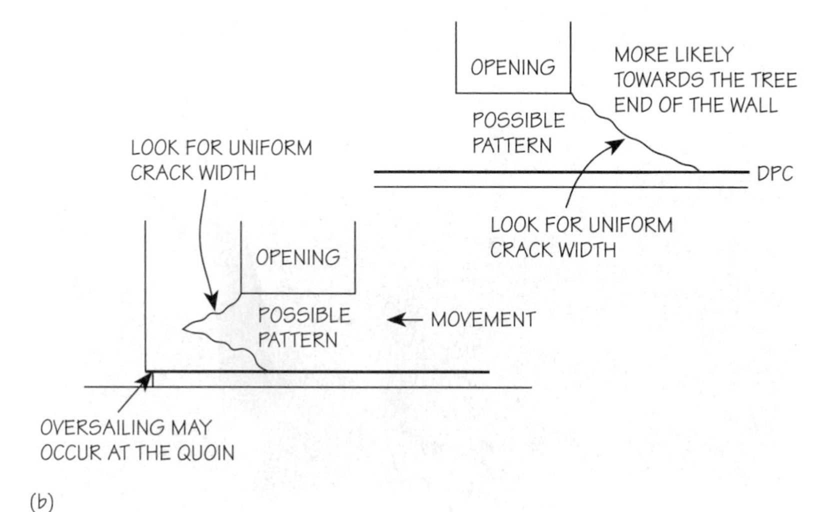


**Fig. 10.9.** Moisture and thermal expansion cracking in clay brickwork: (a) irreversible; (b) reversible. Note that cracking will not usually cross the DPC, and that irreversible and reversible movements may occur coincidentally.

**Fig. 10.10.** Diagonal tensile cracking through mortar joints.

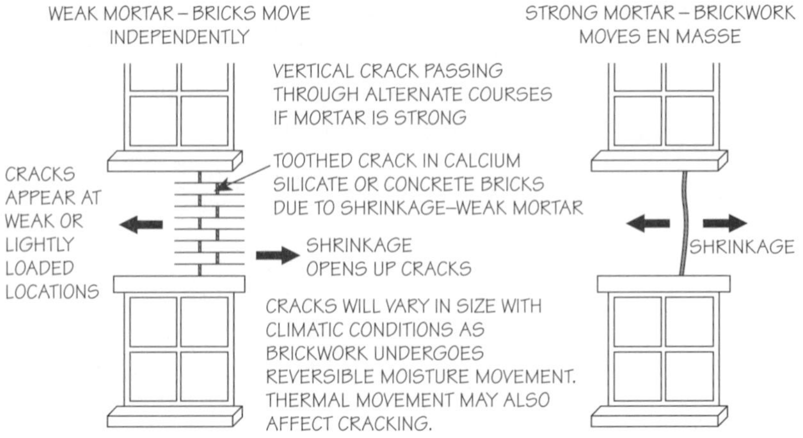


**Fig. 10.11.** Cracking in calcium silicate and concrete bricks. Crack will be of reasonably uniform width. Profiles should match.

# 10.4 Rotational cracking in brick walls

Rotational cracking occurs in walls which have no capacity for rotation, yet are the subject of forces imposed upon them from other, attached walls.

In the simple instance, a wall may rotate under failure of support – caused by movement or failure in a beam, supporting wall or foundation. This can occur at corners of buildings for a variety of reasons, many related to soil weakness. The corner of the building rotates under its self-weight, and perhaps also under imposed loads, producing a crack which widens towards the top. The hinge point or hinge plane is at the base of the wall. Inspection of the wall will indicate rotational movement, and there may also be a corresponding crack in the floor.

An alternative form of rotational cracking occurs in short returns of walling subjected to rotational forces by attachment to offset walls experiencing expansion. Long returns (700 mm +) usually have capacity to rotate and cracking is less likely. As the attached walls expand, the return has too little scope to accommodate the rotational movement and cracks appear in order to release the stresses. These cracks are vertical, and for a cavity or single-leaf wall are usually located at a distance of one-half of a brick thickness from the quoin. Such cracking does not usually cross the DPC since the brickwork below the DPC is more dimensionally stable (it does not dry out and re-wet so dramatically as the above-DPC brickwork). Movement in an elevation which changes the rectangular shape to a parallelogram is called 'lozenging'.

An alternative source of stress relief, which usually occurs in the elevation affected by the movement stresses, is bulging in the wall. This may be accompanied by vertical tensile cracking at the peak of the bulge.

*The Technology of Building Defects.* Dr John Hinks and Dr Geoff Cook.
Published in 1997 by E & FN Spon, 2–6 Boundary Row, London SE1 6HN, UK. ISBN 0 419 19770 2

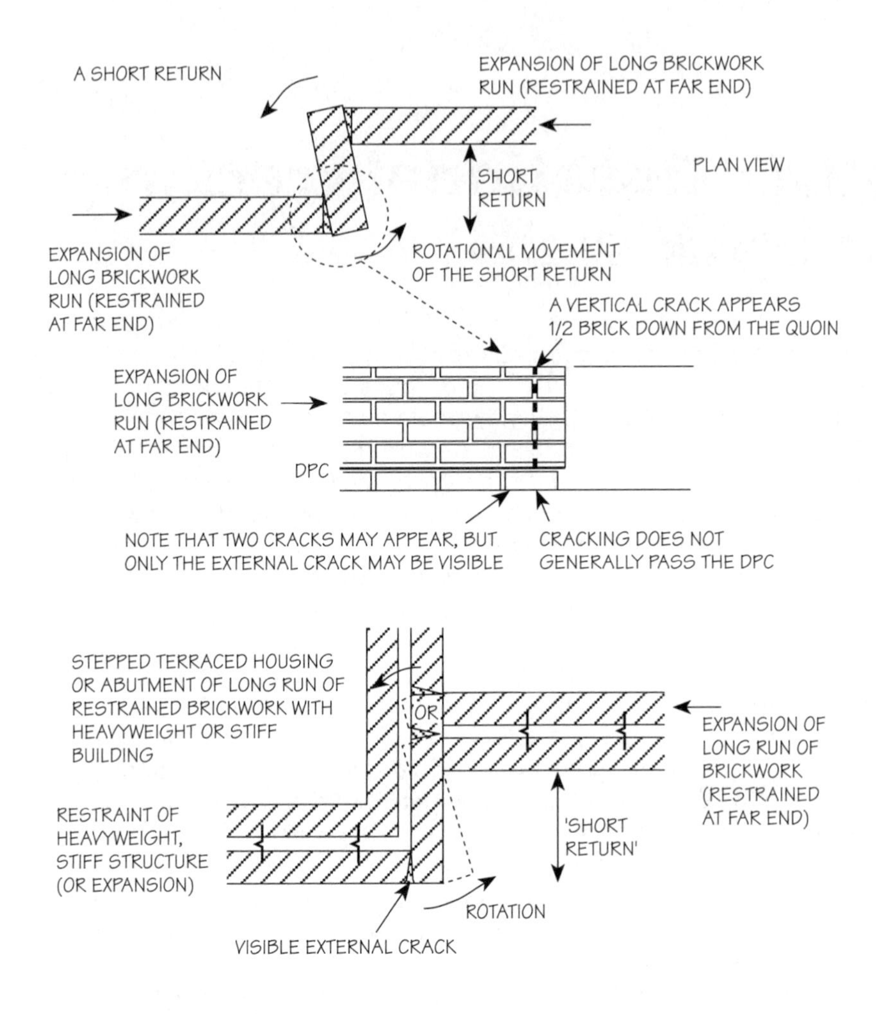


**Fig. 10.12.** Cracking in short returns of brickwork caused by either reversible expansion movement or irreversible expansion of clay bricks.

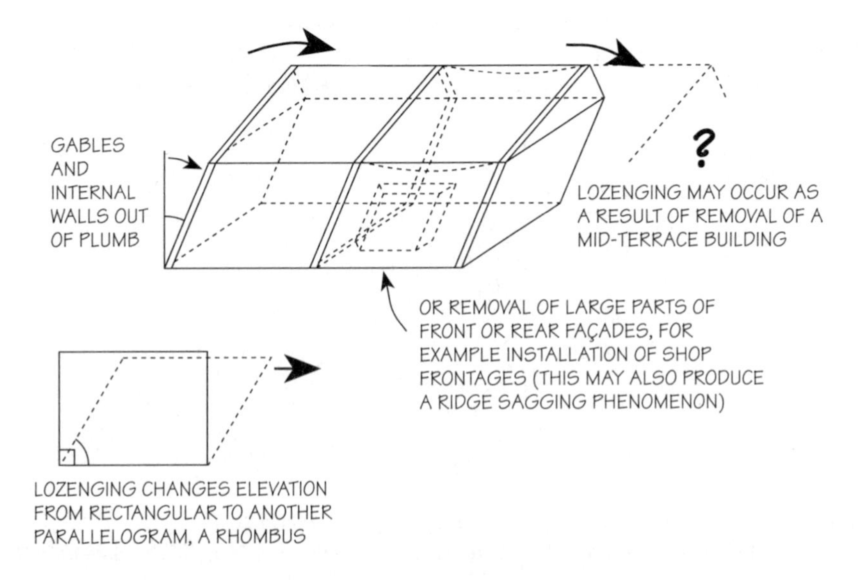


**Fig. 10.13.** Lozenging in buildings.

# 10.5 Oversailing

Oversailing is a phenomenon that occurs at the corners of buildings, usually at the DPC level, where the DPC creates a slip plane and dislocates the wall experiencing stresses from wall below the DPC, or at other locations in the building where a relatively unrestrained element is in contact with a restrained or restraining element – for example, a parapet wall above a concrete roof deck.

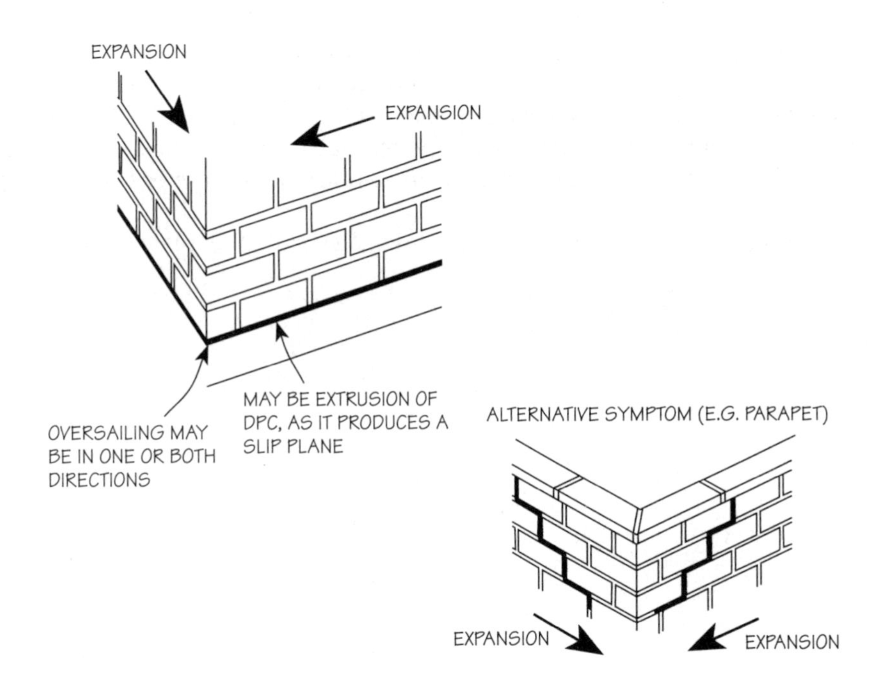


**Fig. 10.14.** Oversailing of walling at corners caused by expansion of long runs of brickwork.

In the former case, expansion in one or both walls meeting at a corner results in the wall slipping across the DPC during expansion, but not withdrawing upon contraction. The forces under expansion are sufficient to push the wall, whereas the forces under contraction, coupled with the relatively low tensile strength of brickwork walls, are insufficient to pull the wall. The defect appears as a progressive overhanging of the wall or parapet in one or two directions. Any DPC in contact with the wall may be extruded also. The extent of overhang is related to the coefficient of expansion and the extent of reversible or irreversible movement. Usual causes are thermal and moisture-related movements.

*The Technology of Building Defects.* Dr John Hinks and Dr Geoff Cook.
Published in 1997 by E & FN Spon, 2–6 Boundary Row, London SE1 6HN, UK. ISBN 0 419 19770 2

# 10.6 Instability in bay windows

Many early bay window constructions were built on little or no foundation and were also poorly connected to the remainder of the frontage. Nevertheless the self-weight of these bays was often significant, especially the two-storey bays.

A consequent defect is the detachment of the bay window, coupled with rotational failure, as it settles in relation to the remainder of the building. Possible symptoms of this include visible cracking at the connection point between the protruding bay window and the remainder of the frontage, which may be coupled with leakage at the head of the bay where it joins the wall. Flashings may have split or been pulled out, allowing rainwater run-off from the façade to enter the building and appear as dampness inside the head of the bay.

An alternative form of the defect is wholesale vertical displacement of the bay.

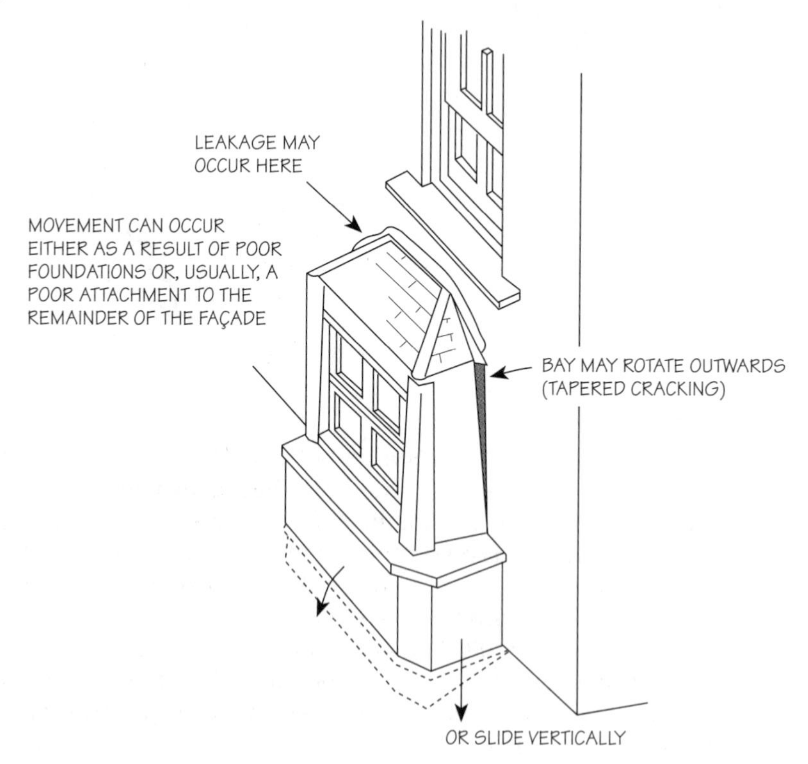


**Fig. 10.15.** Instability in bay window: vertical sliding or rotational settlement with tapered cracking.

*The Technology of Building Defects.* Dr John Hinks and Dr Geoff Cook.
Published in 1997 by E & FN Spon, 2–6 Boundary Row, London SE1 6HN, UK. ISBN 0 419 19770 2

# 10.7 Movement in parapet walls

Movement can occur in parapets for a number of reasons. Instability in parapets is common because of the relative lack of self-weight or restraint which usually limits movement in walls, coupled with a relatively high exposure to the environmental agents of thermal and moisture movement. Parapets are therefore also relatively susceptible to defects in materials, such as frost attack or sulphate attack.

The main mechanisms of movement are irreversible movement, usually expansion in clay bricks (but could be contraction in concrete), and reversible movement such as thermal and/or moisture movement.

The symptoms of parapet movement are usually oversailing at corners, produced during the expansion cycles of thermal and moisture movement. There may be accompanying cracking in the brickwork.

Restraint at the base of the parapet wall, from its attachment to a concrete roof deck, for instance, can produce a degree of dimensional control and therefore stability. However, the roof deck itself may not be dimensionally stable and movement here can also produce symptoms in the parapets. In cases of restraint at the base of the parapet there may still be expansion problems in the exposed, yet relatively unrestrained, head of the parapet. This cracking is similar to hogging failure.

*The Technology of Building Defects.* Dr John Hinks and Dr Geoff Cook.
Published in 1997 by E & FN Spon, 2–6 Boundary Row, London SE1 6HN, UK. ISBN 0 419 19770 2

**Fig. 10.16.** External cracking to corner of parapet wall originating at corner of rainwater opening. Expansive effects evident from increase of crack width with height.

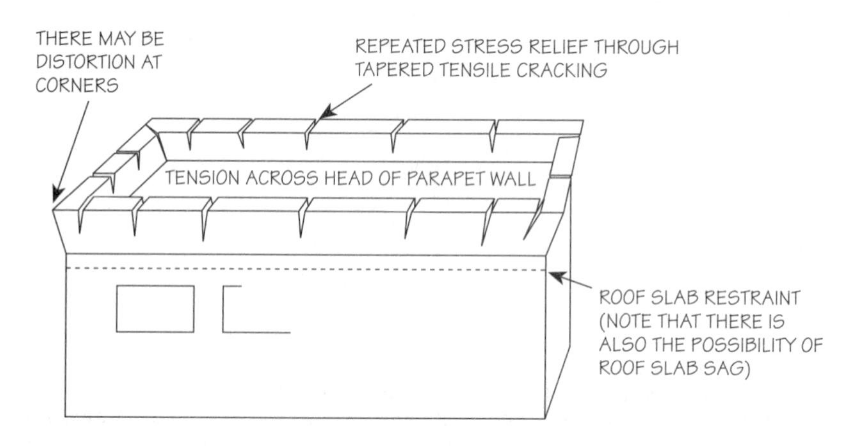


**Fig. 10.17.** Parapet wall tapered tensile cracking with base restraint.

# 10.8 Internal cracking associated with thermal expansion

This defect is not strictly related to walls, but operates using the same mechanism as oversailing, within the plane of a (flat) roof deck or floor slab.

The mechanism usually operates through differential movement in one element of the structure which causes tension at its connections with the remainder of the structure. A possible scenario is that of a flat roof exposed to sunlight and therefore undergoing cyclic thermal movement. In a free-standing and broadly symmetrical building plan and structure the forces will tend to be distributed evenly in all directions. In uneven structures, or plan layout (or alternatively, where the building is attached to a stiff structure), the movement may be concentrated in one direction only (away from the point of restraint). With large elements and/or large temperature (or moisture) changes, the resultant movement may produce significant distortion at connections. This may appear as relief cracking in internal walls or distortion of columns.

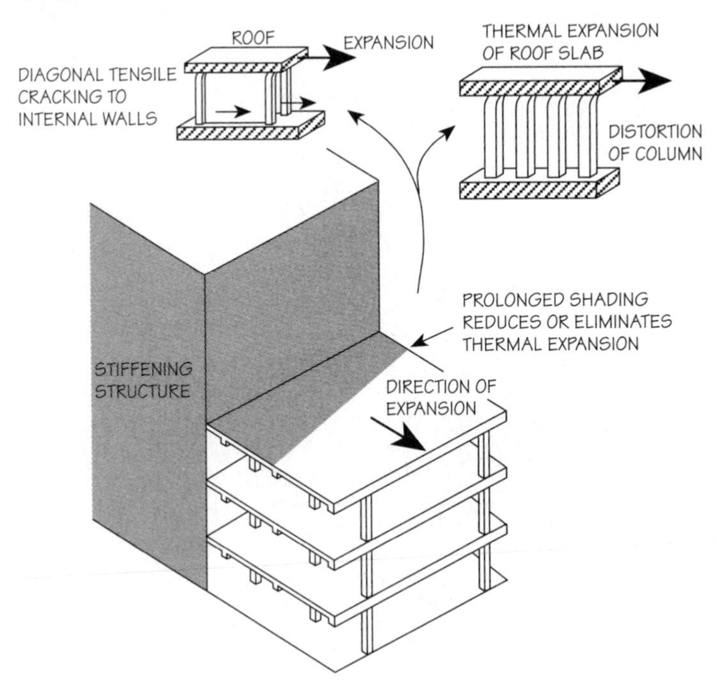


**Fig. 10.18.** Concrete roof slab thermal expansion in framed buildings.

*The Technology of Building Defects.* Dr John Hinks and Dr Geoff Cook.
Published in 1997 by E & FN Spon, 2–6 Boundary Row, London SE1 6HN, UK. ISBN 0 419 19770 2

# 10.9 Tensile cracking failure in concrete frames with brick infills

This defect was characteristic of early concrete-framed buildings which were then built with infill brickwork panels omitting movement joints. This problem is now well understood, but the principle is applicable to a number of scenarios where compressive and tensile forces meet without the scope to accommodate their relative movement. For instance, a series of failures occurred in early cladding and curtain walling for the same reasons (see elsewhere in this book).

As with all building cracking, the cracks take the line of least (tensile) resistance. Where the tensile resistance exceeds the stresses required to

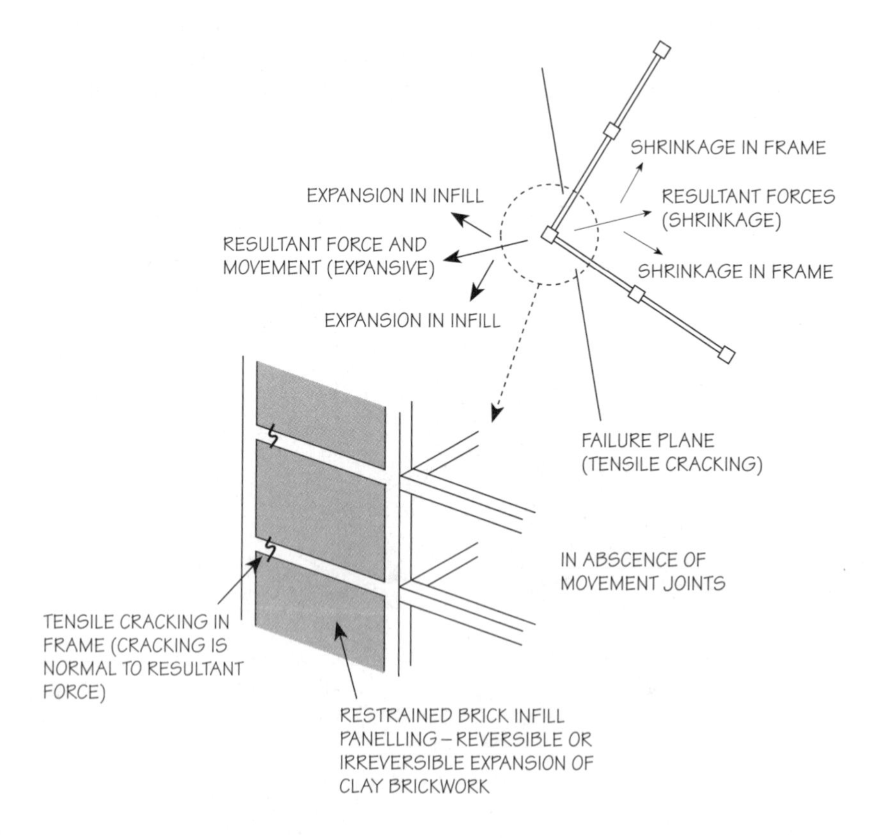


**Fig. 10.19.** General tensile cracking failure to concrete frames with brick infill panels.

*The Technology of Building Defects.* Dr John Hinks and Dr Geoff Cook.
Published in 1997 by E & FN Spon, 2–6 Boundary Row, London SE1 6HN, UK. ISBN 0 419 19770 2

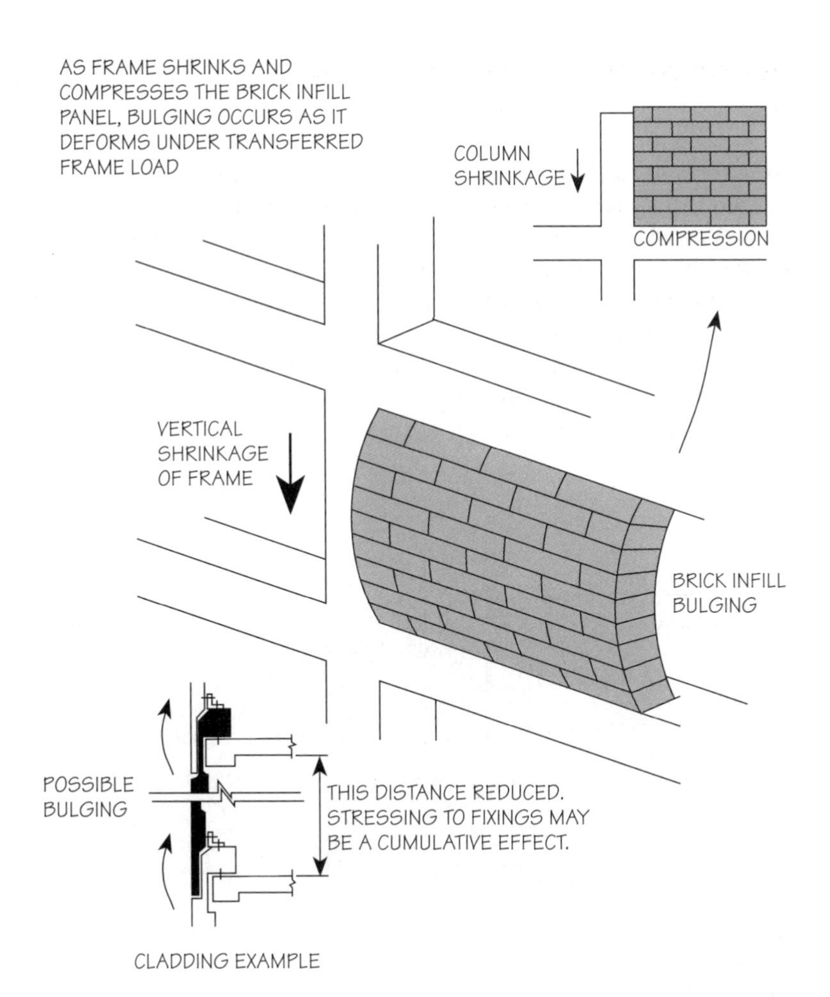


**Fig. 10.20.** Shrinkage of in-situ concrete frames.

induce bulging, as may occur with reinforced concrete, the defect may appear as bulging in the weaker element. The root cause of the defect is the same in each case.

The phenomenon of cracking to concrete frames and/or bulging to restrained brick infill panels works by one of two similar mechanisms. An in-situ concrete frame undergoes shrinkage as it dries out, in addition to compressive shrinkage as the self-weight of the building is imposed during construction. This can produce shrinkage in all directions, but predominantly vertically.

As the frame shrinks it may crack at weak points, usually around corners, as the brickwork resists compression. The problem will be exacerbated by differences in the thermal expansion coefficients of the two materials, and more so by moisture differentials. Whilst the in-situ frame will undergo irreversible shrinkage early in its life (as the water content from the green concrete dries out), in contrast the brickwork may expand as it takes up moisture to harmonize with the surroundings.

This irreversible expansion occurs in the period immediately after firing of the bricks. Bricks which are not used too soon after firing will therefore not cause a particular problem with irreversible shrinkage. The alternative to cracking in the frame is bulging in the brick panels.

## Interaction between brick cladding and concrete frames

Under long-term shrinkage of concrete frames, it is possible for attached or encapsulated brick cladding to be squeezed beyond its design load capacity, thereby causing distortion as a conflict arises between the short- and long-term irreversible shrinkage phenomena (in the concrete frame) and the reversible movement in the brick cladding. This may be exacerbated by cumulative inaccuracies between the frame and cladding – which may have been accommodated during construction by hacking back precast concrete floor units to make them fit, thereby reducing the reinforcement cover and creating other (secondary) defects.

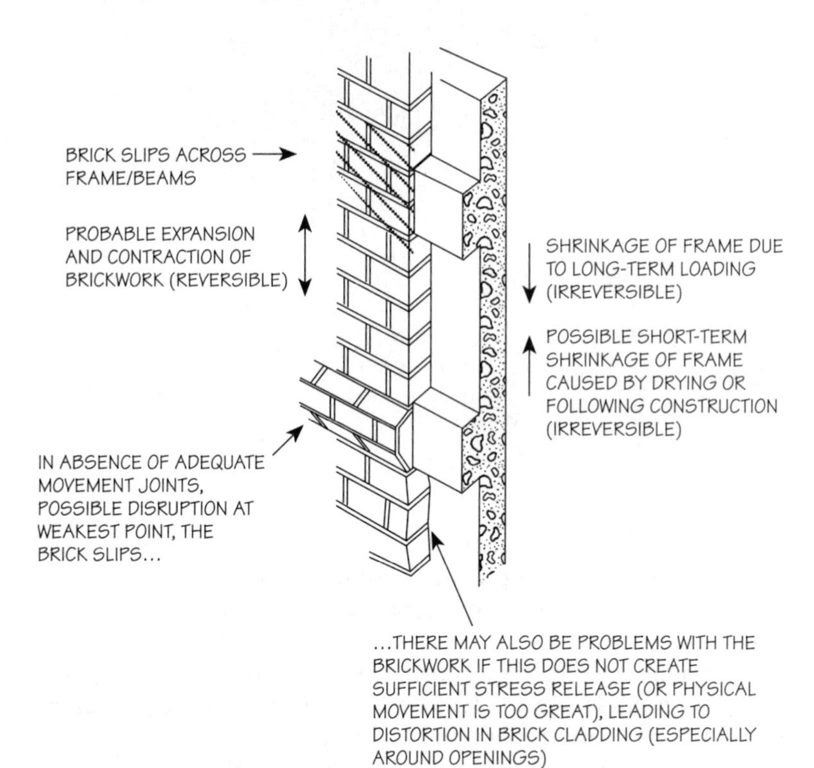


**Fig. 10.21.** Interaction of brickwork enclosed by concrete frame.

# 10.10 Defects in chimneys

In use, chimneys may suffer from sulphate attack brought on from the sulphates in the smoke passing through them. There was also an era when it was fashionable to remove gable chimneys inside houses, leaving the section within the roof and above. This causes its own problems.

**Fig. 10.22.** Failure of chimney stack. Possible sulphate attack of upper section of stack and weathering damage to the brick corbelling.

*The Technology of Building Defects.* Dr John Hinks and Dr Geoff Cook.
Published in 1997 by E & FN Spon, 2–6 Boundary Row, London SE1 6HN, UK. ISBN 0 419 19770 2

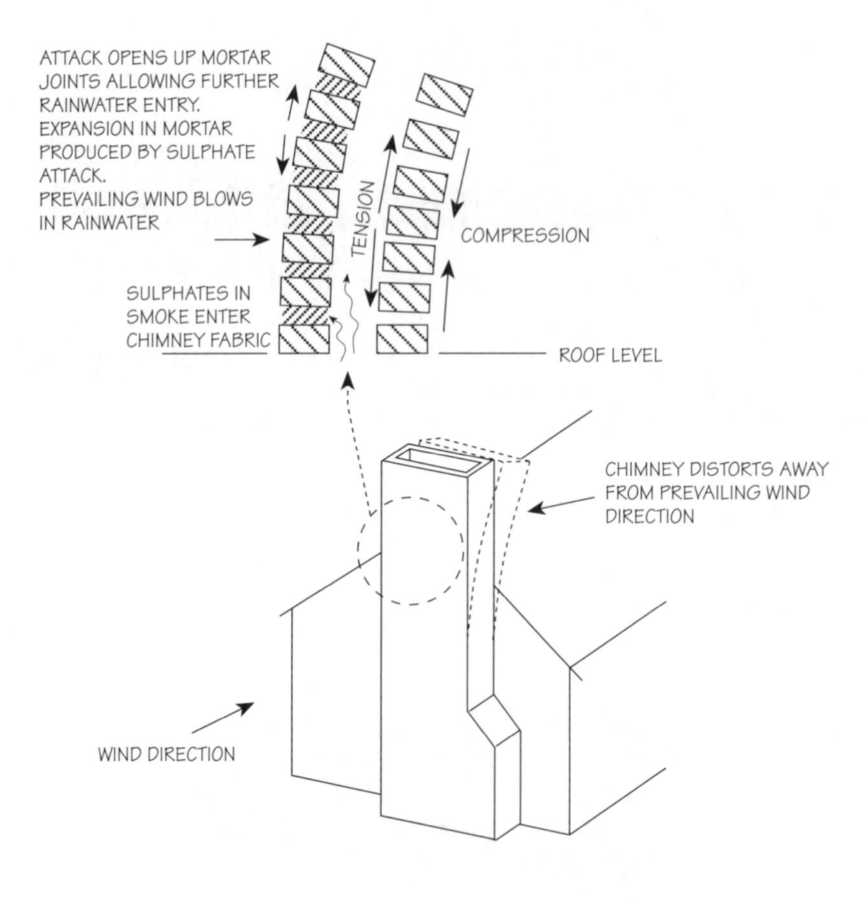


**Fig. 10.23.** Sulphate attack in unlined chimney.

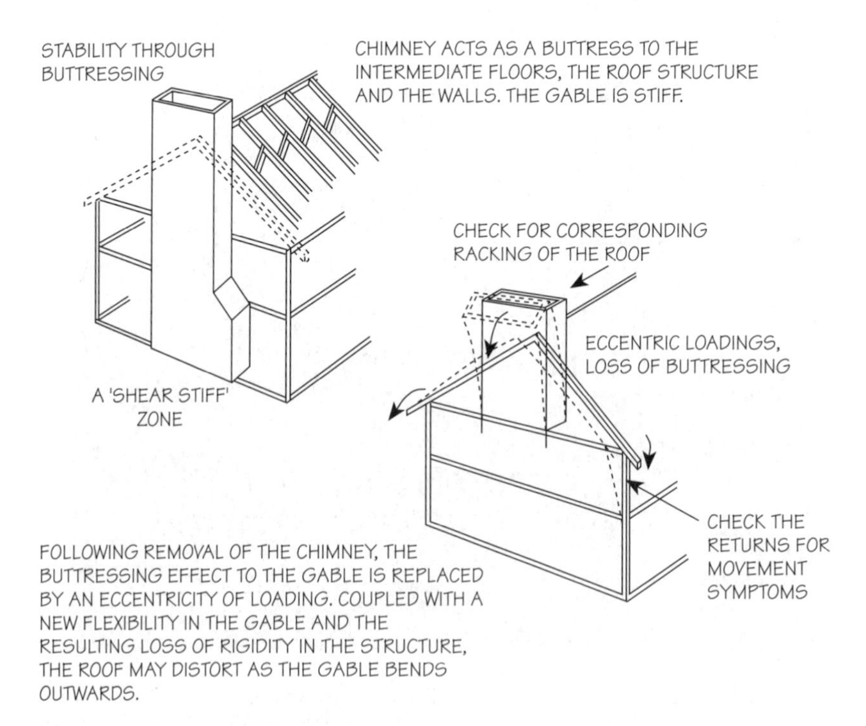


**Fig. 10.24.** The effect of chimney removal on gable walls.

## Sulphate attack in chimneys

Sulphates in smoke passing through early unlined chimneys pass into the fabric of the chimney. The expansive reactions associated with sulphate attack occur in the mortar in the presence of moisture. For chimneys the exposure is usually related to wind direction, being greatest on the side facing into the predominant wind direction. This side is exposed to moisture more than the others and a differential sulphate attack expansion occurs, producing a characteristic bend in the chimney away from the penetrating winds. This may pull over the attached gable. The bending opens up the exposed mortar joints, which accelerates the process.

## Partially removed gable chimneys

The removal of the lower portion of a chimney to make more room inside a house which no longer relies on solid fuel or other flue-dependent heating was an occasional modification to houses some years ago. The effect of this is for the remaining portion of the chimney, which is attached to an open gable, to produce top-heaviness in the gable (which is usually too slender to support it).

The symptom may be deflection in the gable, which will probably produce racking in the roof structure. The chimney and upper gable may obviously lean, leading to cracking and instability.

# 10.11 Further reading: movement and distortion problems in walls generally


Addleson, L. (1992) *Building Failures – Guide to Diagnosis Remedy, and Prevention*, Butterworth-Heinemann.

Alexander, S.J. and Lawson, R.M. (1981) *Design for Movement in Buildings*, Construction Industry Research and Information Association Technical Note 107.

BRE (1979) *Estimation of Thermal and Moisture Movements and Stresses*, Building Research Establishment.

BRE (1987) *External Masonry Walls – Assessing Whether Cracks Indicate Progressive Movement*. Defect Action Sheet DAS 102, Building Research Establishment.

BRE (1988) *Common Defects in Low-Rise Traditional Housing*, Digest 268, Building Research Establishment.

BRE (1989) *Simple Measuring and Monitoring of Movement in Low-Rise Buildings Part 2: Settlement, Hecure and Out-of-Plumb*, Digest 344, Building Research Establishment.

BRE (1990) *Cracks in Buildings Due to Drought*, Leaflet XLI (November), Building Research Establishment.

BRE (1990) *Assessing Traditional Housing for Rehabilitation*, Report 167, Building Research Establishment.

BRE (1990) *Surveyors Checklist for Rehabilitation of Traditional Housing*, Report 168, Building Research Establishment.

BRE (1991) *Repairing Brick and Block Masonry*, Digest 359, Building Research Establishment.

BRE (1991) *Why Do Buildings Crack*? Digest 361, Building Research Establishment.

Bryan, A.J. (1989) *Movements in Buildings*, Report R3, Chartered Institute of Building Technical Information Service.

CIRIA (1986) *Movement and Cracking in Long Masonry Walls*, Construction Industry Research and Information Association Special Publication 44.

HMSO (1989) *Defects in Buildings*, Property Services Agency.

ISE (1980) *Appraisal of Existing Structures*, Institution of Structural Engineers.

Launden, B.S. (1988) *Remedying Defects in Older Buildings*, Chartered Institute of Building Technical Information Service.

Macgregor, J. (1985) *Outward Leaning Walls*, Technical Pamphlet 1, Society for the Protection of Ancient Buildings.

Ranger, P. (1983) *Movement Control in the Fabric of Buildings*, Mitchell's Series, Batsford Academic and Institutional.



*The Technology of Building Defects.* Dr John Hinks and Dr Geoff Cook.
Published in 1997 by E & FN Spon, 2–6 Boundary Row, London SE1 6HN, UK. ISBN 0 419 19770 2

CHAPTER 11

# Defects in floors

## Learning objectives

You should understand:

- the range of structural defects in floors, and their consequence for the floor and the remainder of the building;
- the problems with dampness in floors.

Floors exhibit a range of defects related to their expected performance in use. The main functions of flooring are structural adequacy, including the transfer of all loads (dead and imposed) to the ground or the walls/ foundations without deflecting excessively; to resist water penetration; to control thermal losses; and to provide a safe and stable surface. Special requirements include fire resistance and control of acoustic transmission. Floors can be deficient in any or all of these contexts. There may also be problems with the floor finish, dealt with separately.

## Structural inadequacy in floors

Floors may be structurally inadequate because of failures in their bearing at their end supports (or in the case of large spans, possibly at the mid-support also). They may also be defective through excessive span for their sizing, or overloading beyond design specification. Excessive notching, especially in the sensitive middle third of the span, can produce bending as a result of overloading.

The end support/connections for timber flooring joists can be achieved in a number of ways, and deficiency will produce a range of characteristic symptoms including sagging of the floor at edges or particular weak locations, springiness (which may be due to displaced or rotating joist hangers) and local collapse, perhaps due to rot in the bearing ends of the joists or local failure of the support wall.

These problems will be exacerbated where they occur in conjunction with either of two other faults affecting structural stability – the absence of strutting and/or lateral bracing or packing to that bracing (which should be connected to parallel walls and span at least two joints). In the latter case there may be consequential instability in a wall which is designed to derive support from the floor. In upper floors the attendant symptoms of perimeter support problems with floors may include ceiling cracking, or perimeter sagging and gaps in ceilings.

Overloading (directly or as a secondary result of deterioration or retrospective notching of floor timbers) can be expected to produce sagging,

*The Technology of Building Defects.* Dr John Hinks and Dr Geoff Cook.
Published in 1997 by E & FN Spon, 2–6 Boundary Row, London SE1 6HN, UK. ISBN 0 419 19770 2

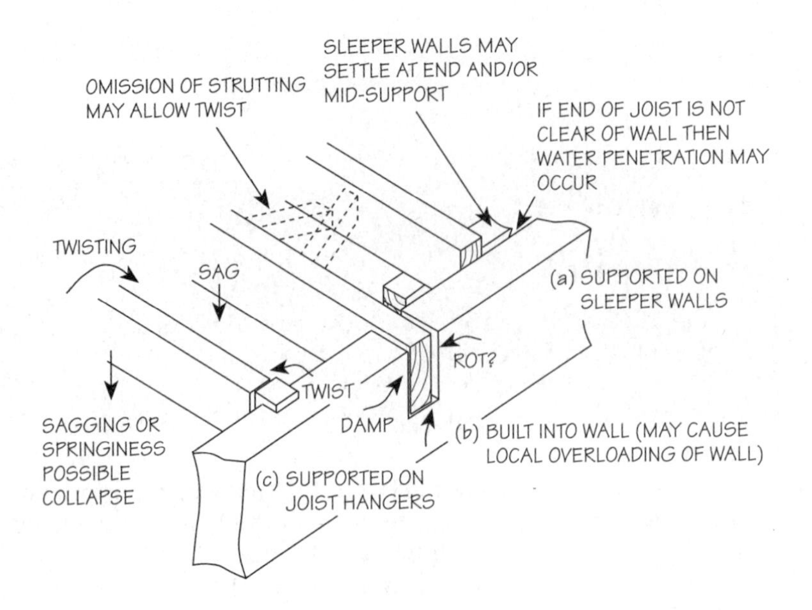


**Fig. 11.1.** Composite schematic diagram of bearing problems in flooring. Sagging may be caused by overloading, or collapse/dislocation of end bearing.

which may in turn produce ceiling cracking to suspended floors. Note, however, that ceiling cracking may be a symptom of movement/detachment of plasterboard (usually regular patterns). More relevant symptoms include jamming doors and gaps above partitious (also 'ringing' of partitions when struck, which suggests that they are carrying loads). Distortion in floors may be a direct fault in the floor, or it may be a symptomatic response to movement in the remainder of the structure, such as may be caused in various forms by settlement, subsidence, ballooning, lozenging and book-ending. The symptoms will vary with the nature of the primary cause and the structural character of the floor.

In concrete suspended floors there may be problems with connectivity between the concrete and shear connections to the supporting beams, and also with inaccurately placed or incorrectly developed reinforcement.

There may be similar inadequacies of bearing at walls, as with timber floors, and localized failure of supports, which can produce sagging at the perimeter of the floor. There is also the possibility of cold bridging with such details if the floor crosses the cavity or is supported in a solid wall.

In upper floors, generally, there can be severe structural consequences of the installation of services using notching, especially where this has been accompanied by the movement of strutting.

In ground floors, particular defects include moisture movement caused by variations in the soil moisture content (where the floor is not isolated by a DPM); by water penetration of a faulty DPM or DPM/DPC connection, or by water entering via an adjacent wall. As with all cracking, the symptoms appear at the weak points, such as in the screeding, as ripples, or around the perimeter.

This type of fault may aid sulphate attack on unprotected concrete if the soil conditions are suitably aggressive. Cracking and differential movement in solid ground floors may also be the result of instabilities in the surface of the soil contacting the floor. Examples of this mechanism include clay soil expansion (or relative stability as the surrounding soil shrinks in dry conditions, leaving the relatively isolated central plug proud of the new ground level) or shrinkage/settlement of fill.

Suspended concrete ground floors may be uneven owing to failures of inadequate bearing, unevenness in bearings or irregularities in block infilling. This may include localized overloading of the blocks.

## Dampness in floors

Dampness in flooring can produce a range of direct problems in the form of localized rot (dry or wet rot). If unattended, this rot can become extensive, especially where a lack of ventilation due to blockage or omission of air bricks and cross-ventilation occurs. This can cause particular problems where isolated solid floors create zones of dead air in abutting timber flooring. It is important to be able to identify the sources of the damp. Common sources include flooding of underfloor voids via air bricks or breaches in DPMs, DPCs and/or their connections. Direct water penetration may also occur with plumbing or drainage (which may have been damaged by another movement-related defect).

Rot may extend to sole plates and floor coverings, which with chipboard flooring can be quite destructive.

Chipboard can exhibit a range of defects. It must be well supported and fixed, and of the correct (flooring) grade with appropriate perimeter detailing. The consequences of inattention to fixing details can include sagging or buckling or loose, squeaking boards.

Floors frequently exhibit secondary defects in buildings suffering other problems. They can be a vital stiffening component in a building, however, and must be considered as an important item during inspection.

### Revision notes

- Problems arise with structural adequacy, deflection in use, damp penetration, thermal and acoustic insulation and fire resistance.
- Possible mechanisms of structural deficiency include the following.

#### Solid ground floors

- Shrinkage or settlement of supporting soil or fill.
- Surface cracking due to omission of top reinforcement.
- Moisture movement in slab, which may produce rippling in (thinner, weaker) screed.
- Localized water penetration with failed DPM, DPC or DPM/DPC connection, which can lead to sulphate attack of unprotected concrete in contact with aggressive soils.

## Suspended ground floors

- Inadequacies in bearing support, including local failure or unevenness in support.
- Damp penetration of joist ends in walls or encroaching into cavities.
- otational failure of joist hangers.
- Sagging of sleeper walls.
- Sagging of floors due to overloading or other progressive failure.
- Omission of strutting.

## Upper floors

- Inadequacies on bearing support.
- Damp penetration of built-in ends of joists.
- Rotational failure of joist hangers.
- Overloading.
- Weakening of floors by notching for services.
- Removal/omission of strutting, lateral bracing.
- Dampness may also be a problem; caused by:
  - water crossing DPC;
  - water bridging omitted or defective DPM or DPC/DPM connection;
  - air bricks allowing water entry;
  - direct water penetration (plumbing, drainage leaks).
- Consequent problems include dry and wet rot of timbers.
- Damp may affect sole plates and floor coverings also.
- All of these may be compounded by lack of ventilation.
- Lack of ventilation may be exacerbated by omission, blockage or misalignment of air paths, creating zones of dead air.
- Floors (especially upper floors) contribute to structural stiffness of many buildings and faults in their structure may have consequences for the building as a whole.
- Chipboard flooring may exhibit sagging, or buckling where it is inadequately fixed or supported. Loss of strength can occur if exposed to dampness or high humidity.

## Discussion topics

- Describe the range of causes of dampness in floors and the possible consequences for the building as a whole of unchecked dampness.
- Explain how floors contribute structurally to buildings, and what the various types of defect in floor structure can mean for a building. Discuss the possible symptoms of structural defects in floors, comparing those on ground and upper floors.

## Further reading

Addleson, L. (1992) *Building Failures: Guide to Diagnosis, Remedy, and Prevention*, Butterworth-Heinemann.
BRE (1966) *Moisture of Construction as a Cause of Decay in Suspended Ground Floors*, Technical Note 14, Building Research Establishment.
BRE (1983) *Suspended Timber Floors – Chipboard Flooring – Specification*, Defect Action Sheet DAS 31, Building Research Establishment.
BRE (1984) *Suspended Timber Floors – Joist Hangers in Masonry Walls – Specification*, Defect Action Sheet DAS 57, Building Research Establishment.
BRE (1986) *Suspended Timber Floors – Remedying Dampness Due to Inadequate Ventilation*, Defect Action Sheet DAS 73, Building Research Establishment.
BRE (1986) *Suspended Timber Floors – Repairing Rotted Joists*, Defect Action Sheet DAS 74, Building Research Establishment.
BRE (1986) *Suspended Timber Floors – Notching and Drilling of Joists*, Defect Action Sheet DAS 99, Building Research Establishment.
BRE (1987) *Wood Floors – Reducing the Risk of Recurrent Dry Rot*, Report 103, Building Research Establishment.
BRE (1990) *Timber ground floors*, in *Assessing Traditional Housing for Rehabilitation*, Report 167, Building Research Establishment, ch. 4.
BRE (1990) Timber intermediate floors, *Assessing Traditional Housing for Rehabilitation*, Report 167, Building Research Establishment, ch. 6.
Dove, E. (1983) *Suspended Concrete Ground Floors for Houses*, British Cement Association.

CHAPTER 12

# Problems with internal finishes

## Learning objectives

- You should be able to identify the symptoms and causes of a range of defects in a range of internal finishes.
- This will include an understanding of the role of workmanship and surface preparation.
- It is important that you are aware of the influence of differential movement in the deterioration of surface finishes.

## Screed

Screed requires a high standard of workmanship to produce a durable finish. Excessive trowelling can produce surface crazing.

Large areas of screed laid in one operation can cause random cracking due to differential shrinkage. Alternatively, laying screeds in bays can cause the edges to curl. This may show through thin finishes. A rapid curing time for screed can result in considerable shrinkage and a general loss of strength of the screed.

Where screed is not laid over a DPM there is a risk of dampness problems.

An inadequate bond between the screed and the base concrete is a common failure mode. This may be due to the following.

- Poor mix design of screed. In general, mix proportions of > 1:3 cement to aggregate or < 1:4.5 may result in differential movement between the screed and the base concrete.
- Excessive water. This may be added to improve the workability of the mix. The high surface area of the small-sized aggregate can tend to reduce workability. Although adding more water will increase shrinkage it may also cause laitance to form on the surface of the screed, causing dusting.
- Too much time between casting base concrete and laying screed and a poor texture of concrete base. These factors are related to the time between casting and laying the screed. The base concrete may not be swept clean of site debris. Where the time between laying concrete and screed to get monolithic construction is > 3 hours then the screed and concrete may shrink differentially. Where the screed is laid within 3 hours of the base concrete but exceeds 25 mm thickness or contains aggregate > 10 mm then differential shrinkage can occur. Where the screed is laid on mature concrete it may fail if there is inadequate mechanical key, formed by hacking or scarifying, with the concrete base.

*The Technology of Building Defects.* Dr John Hinks and Dr Geoff Cook.
Published in 1997 by E & FN Spon, 2–6 Boundary Row, London SE1 6HN, UK. ISBN 0 419 19770 2

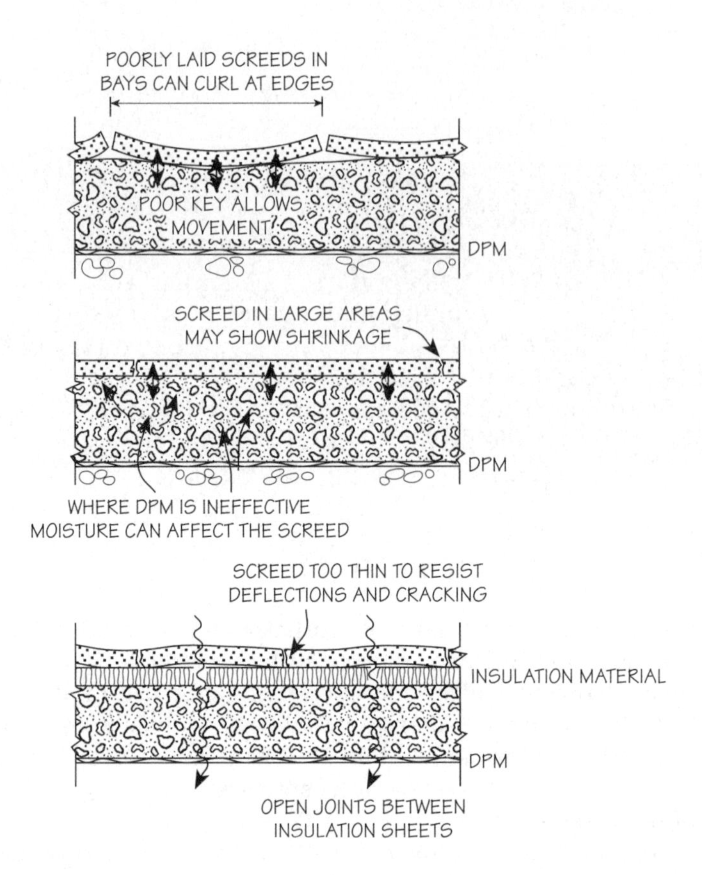


**Fig. 12.1.** Screeds – typical defects.

Surface laitance may not have been removed or the surface adequately wetted before laying the screed. There may be bonding problems with screed < 40 mm thick.

Screeds laid over a DPM may fail where they are < 50 mm thick. These thicker screeds require mechanical compaction to achieve adequate durability. All screeds > 25 mm thick should ideally be laid in two layers immediately after each other. Where this does not occur there may be similar bonding problems to those between the screed and the base concrete.

A floating screed, which can be laid over insulation material, may fill any gaps between insulation material. This may provide impact sound transmission paths and reduce the performance of the floor. Where floating screeds are < 65 mm thick they may have insufficient strength and crack. Screeds < 75 mm thick which contain heating elements may also crack and have unacceptably high surface temperatures. A failure to provide wire mesh netting over the insulating material may increase the risk of cracking.

## Terrazzo

Terrazzo is normally laid onto screed. It may fail because of bonding in the same ways as described for screed. Cracking is commonly due to poor curing over a short time period. Shrinkage effects may become concentrated and the rapid drying-out of the mix may mean that long-term strength is reduced. Surface crazing may be due to excessive cement content in the terrazzo. Powdering of terrazzo finishes may be due to wet mixes; this can cause laitance and produce permanently weak mixes.

## Clay floor tiles

Differential movement between the tile and the base material can be a problem. Shock waves, due to impact loading passing through the floor, may detach tiles.

Thermal expansion coefficient differences may exist. When the floor becomes cold the screed may contract more then the clay tiles and they can become detached.

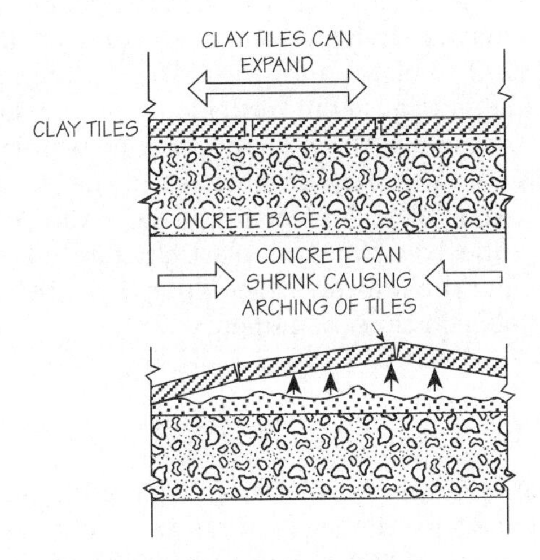


Fig. 12.2. Clay tiles – typical defects.

Differential moisture movement may also occur. Fresh concrete shrinks as it hydrates and becomes dry, whilst the clay tiles may go through an initial, irreversible expansion as they take up moisture from the air. The tiles may take up moisture from the screed or concrete where the DPM has failed. Expansion of the tiles may cause arching. Tiles bedded in mortar on building paper or bitumen felt, to allow for this differential movement to be accommodated, may fail where the concrete floor can key to the building paper. Fresh concrete is more likely to achieve this bond.

Where movement accommodation is not provided around the perimeter, or by dividing the area into bays, failure can occur.

Whilst the chemical resistance of clay tiles is generally good, the mortar joint is less durable. Harsh chemicals in the cleaning water can attack the mortar joints.

## Clay wall tiles

These may fail owing to poor bonding with the substrate. Where a large area of adhesive is applied, this may set before the wall tiles are applied.

Generally the tiles do not move as much as the backgrounds. Tiling over dense mortar or rendered backgrounds on cast concrete may fail since the concrete shrinks and the tiles do not. Dense joints between the tiles mean that any movement cannot be accommodated within the tiled mass. This tends to concentrate movement at corners, demanding the provision of movement joints. Since movement can occur in two directions, vertically and horizontally, then the tiling may fail if there are inadequate movement joints in both directions. Large areas of tiling are at particular risk.

## PVC tiles

The moisture from an insufficiently cured screed, a failed DPM or a failed DPC/DPM junction can cause loss of tile adhesion. The moisture, being alkaline, attacks the adhesive used to stick down the tiles. Tiles with adhesive attached may become detached from the concrete, some curling at the edges. The tiles may be stained from failed adhesive or sodium carbonate following evaporation of the moisture. The sodium carbonate can be absorbed into the tile and cause embrittlement and cracking. The adhesive may fail owing to excessive use of general cleaning water.

Under dry conditions tiles may lift because of the delay between laying the adhesive and laying tiles. Adhesion can be reduced as the adhesive starts to set. Some adhesives allow the plasticizer from tiles to migrate into the adhesive, softening the adhesive. The floor tiles are then able to move around and joints between them can open.

## Timber flooring

There are, of course, a range of insect and fungal problems associated with timber. Many problems with timber flooring can occur owing to the timber being, or becoming, damp. Where boarding is fixed directly to a ground floor concrete slab, typically by the use of timber battens, then failure of the DPM or failure of the junction between the DPM and DPC can allow moisture to migrate through the slab.

Moisture may also come from the screed where there is insufficient time for the screed to cure and dry out. A general recommendation is 1 month for

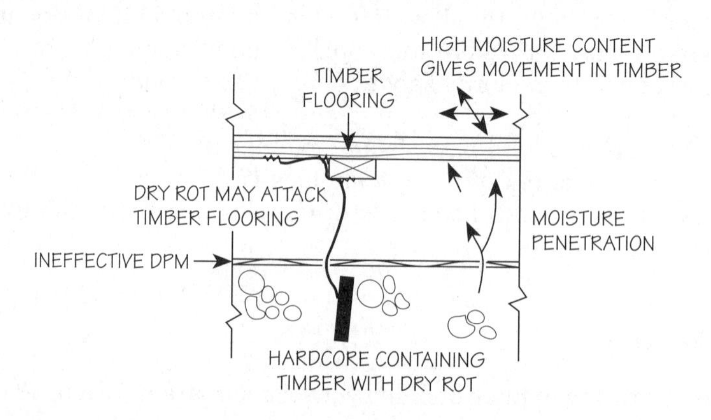


**Fig. 12.3.** Timber flooring – typical defects.

every 25 mm of screed thickness. It is possible for dry rot from infected timber in the hardcore to move through the concrete and screed and attack the new timber finish.

Where moisture movement is not accommodated then expansion can cause arching and lifting. This is a particular problem with wood block flooring. Problems may occur when wood blocks are laid at a moisture content which they would not experience in service. Then when on site to condition to the general humidity, timber may absorb water to above that required in service. When the timber is subsequently fixed it may shrink. Using a high-moisture-movement timber will increase the potential for movement.

Movement joints, usually compression, provided around the wall can accommodate this movement. The adhesive between block and mortar base may fail or the bond between mortar base and the concrete.

Floorboards laid with the heartwood downwards can produce convex surfaces and curled edges. This can also cause a pattern when thin coverings are laid over the boards. Chipboard flooring can expand irreversibly when wetted. There can also be substantial strength loss. Poor fixing or inadequate chipboard thickness for the span can cause excessive deflection or cracking.

## Plaster

Traditional lime-and-sand plaster ceilings generally cracked. Where laths were used these can become brittle with age and no longer accommodate movement.

An uneven finish generally indicates poor workmanship. Inadequate mixing and the addition of excessive amounts of water may produce powdery and weak surfaces.

Plaster can fail where the incorrect plaster has been applied to the background or there is insufficient mechanical key. The browning or undercoat should be matched to the suction and mechanical key provided by the background. The finish coat must then be matched to the undercoat or browning. Within the context of failures in plaster, the failure to match plasters and backgrounds is relatively common. The bulging of large areas of plaster, perhaps with associated cracking, may be caused by a failure of the plaster to adhere to the substrate. This may sound hollow.

A failure to wet high-suction backgrounds can reduce the bonding between plaster and background and may reduce the strength and durability of the plaster. A plaster finish coat can be applied to certain types of plasterboard. Boards commonly have two surfaces and application of the finish coat to the incorrect side results in inadequate bond and failure.

Because gypsum plaster is very sensitive to dampness, it may fail in locations where the background has a high moisture content, e.g. walls above a newly inserted DPC or in defective basements.

Moist plaster can cause corrosion of ferrous metals within it. This is due to the relatively acidic nature compared to cement, of gypsum plaster.

Cracking of plaster may be associated with the thermal and moisture movement of the background. They may also be due to structural movement. Differential movement can cause cracking at internal corners where a plasterboard ceiling meets an in-situ plaster wall. In-situ plaster on concrete block walls can fail where the blocks were damp when the plaster was applied. The concrete blockwork will shrink when drying out and the plaster will expand when drying out, causing the plaster to become detached. Where a cementitious undercoat is applied, this may move with the blockwork, causing only the finishing coat to become detached.

Plasterboard ceilings can crack along the joints of boards, particularly where the joints are not correctly made. Sagging boards may be related to inadequate fixings, particularly a lack of noggins at plasterboard edges.

## Revision notes

- Screeds and terrazzo require a high standard of workmanship. Differential shrinkage cracking can occur owing to curing and drying effects. May fail to bond to the background because of poor key, thickness or mix compatibility.
- Differential thermal or moisture movement and inadequate accommodation of movement can cause detachment of clay floor and wall tiles.
- PVC tiles may fail because of rising alkaline dampness. The plasticizer from some tiles can soften tile adhesive.
- Timber flooring can become defective owing to fungal or insect attack. The moisture required can come from dampness or the construction. Poorly matching the moisture content of timber during installation with that expected in normal service can cause dimensional changes.
- Inadequate mechanical key to the background, or incompatibility with the background, can cause plaster to become detached. Detached areas may bulge and crack. Gypsum is sensitive to dampness and may induce the corrosion of ferrous metals. Cracking can be due to differential movement.

## Discussion topics

- 'Laying cement floor screeds is a common and straightforward construction activity which rarely suffers from any serious defects'. Discuss.
- Compare the failure mechanisms of clay and PVC tile flooring.
- Explain the differential movement characteristics of timber board flooring, chipboard and wood block flooring.
- Describe a method for the accommodation of movement of an area of glazed wall tiling.
- Identify a range of defects associated with plaster which could be classified as workmanship-related or design-related.

## Further reading


BRE (1963) *Sheet and Tile Flooring made from Thermoplastic Binders*, Digest 33, Building Research Establishment, HMSO.

BRE (1983) *Suspended Timber Floors: Chipboard Flooring – Storage and Installation (Site)*, Defect Action Sheet 32, Building Research Establishment, HMSO.

BRE (1984) *Floors: Cement-based Screeds – Specification (Design)*, Defect Action Sheet 51, Building Research Establishment, HMSO.

BRE (1984) *Floors: Cement-based Screeds – Mixing and Laying*, Defect Action Sheet 52, Building Research Establishment, HMSO.

BRE (1989) *Internal Walls: Ceramic Wall Tiles – Loss of Adhesion*, Defect Action Sheet 137, Building Research Establishment, HMSO.

BRE (1991) *Replacing Failed Plaster*, Good Building Guide 7, Building Research Establishment, HMSO.

BRE (1991) *Design of Timber Floors to Prevent Decay*, Digest 364, Building Research Establishment, HMSO.

Cook, G.K. and Hinks, A.J. (1992) *Appraising Building Defects: Perspectives on Stability and Hygrothermal Performance*, Longman Scientific & Technical, London.

PSA (1989) *Defects in Buildings*, HMSO.

Ransom, W.H. (1981) *Building Failures: Diagnosis and Avoidance*, 2nd edn, E. & F.N. Spon, London.

Richardson, B.A. (1991) *Defects and Deterioration in Buildings*, E. & F.N. Spon, London.

Staveley, H.S. and Glover, P. (1990) *Building Surveys*, 2nd edn, Butterworth-Heinemann, London.

CHAPTER 13

# Defects in roofs

# 13.1 Felted flat roofs

## Learning objectives

- You should understand how the symptomatic failure of felted roofs relates to material, component and (interconnected) elemental failure.
- It is important to understand fully what has been one of the most mythologized buildings defects.

Roofs are among the most difficult parts of a building to survey and unless access to adjacent properties allows a view from above, the inspection of flat roofs is particularly difficult.

These difficulties lead to poor maintenance, with inspections rarely carried out at the recommended intervals of two years. Consequently, flat roofs can require extensive repair work to major failures.

## Defects with felt roofs

The defects associated with domestic flat roofs are interrelated and some of the more common faults with built-up covering are summarized in Table 13.1.

Leaking felt layers are a common defect in domestic properties, discontinuity in the covering arising as a result of movement or inadequate attention to lapping and join formation. These are areas where workmanship is critical and should be subject to thorough supervision and inspection, to the extent of water-testing.

## Temperature-induced defects

The main cause of deterioration in the waterproof membrane is temperature. Exposed roof surfaces are subject to a wide temperature range due mainly to sunlight. Degradation of many of these coverings is accelerated by ultraviolet light, which hardens them. This process is compounded by oxidation. Embrittlement occurs at extremes of temperature, owing to evaporation of volatiles at high temperatures and hardening of the covering at low temperatures.

*The Technology of Building Defects.* Dr John Hinks and Dr Geoff Cook.
Published in 1997 by E & FN Spon, 2–6 Boundary Row, London SE1 6HN, UK. ISBN 0 419 19770 2

**Table 13.1** Defects associated with felted flat roofs

| Defect | Cause | Remedy |
|---|---|---|
| Cracking | Differential movement<br>Joints in substrate<br>Lack of consistent support<br>Severe changes of direction | If extensive re-lay, cut, and bond in a patch<br>Provide new sub-base<br>Adopt modified details as BSCP 6229 |
| Small blisters | Expansion of volatile fractions in bitumen can occur with self-finished or mineral felt | If serious apply cut-back bitumen and grit |
| Large blisters | Expansion of trapped air and/ or moisture (may occur between felt layers or between felt layers and the decking) | Cut out and patch.<br>If persistent, consider ventilation or heavy dressings |
| Ridging and cockles | Moisture movement of finish, sub-base or structure | Normal treatment is to relay<br>Site dressings may conceal potential defects |
| Lap or joint failure | Poor bonding<br>Severe tensile stress caused by blisters or ridging | Cut out and re-bond a patch<br>Simple re-bonding may be possible |
| Ponding | Inadequate falls or blocked drainage ways and areas | Remove and re-lay to falls<br>Additional layers of built-up covering may be sufficient |
| General deterioration | Incorrect choice of covering<br>Lack of surface dressing<br>Poor/inadequate falls<br>Lack of regular inspection | Re-lay if deterioration is severe<br>Extensive site dressing can be used to restore exposed surface |
| Loss of chippings | Severe falls<br>Wind erosion of poorly bedded chippings<br>Inadequate depth of chippings | Consider alternative protective measures<br>Select larger-sized chippings |
| Dents and rips | Physical damage | Patch, paying attention to bonding<br>Area may be subject to traffic |
| Sticky or semi-liquid surface | Chemical damage | Remove source of contamination<br>Remove affected covering, patch or re-lay |
| Water ingress | Joint failure<br>Inadequate or defective skirtings<br>Incorrectly weathered joints<br>Failure where pipes, etc penetrate the felt<br>Physical damage | Mechanical fixing around pipes, rainwater outlets and skirtings may be required |
| Splitting parallel to metal eaves trim | Poor fixing of eaves trim | Remove trim and replace with welted felt drip as BPCP 6229 |

Movement stresses may be induced because of the generally high coefficients of linear expansion of bitumen-based materials. They expand and contract a lot as their temperature goes up and down. Glass-fibre and polyester felts generally have a lower expansion coefficient, and so have less moisture movement problems. These temperature-related stresses act two-dimensionally, producing surface crazing. Even with moderate roof areas, this can be critical if suitable control measures are not adopted (particularly around junctions with surrounding structures and at projections).

To overcome temperature-induced defects, surface dressings can be incorporated to improve the solar resistance of roof coverings and limit the temperature movement at the surface. These dressings include stone chippings, solar-reflective paints and pre-finished coverings.

Some problems have emerged with solar coatings, however, mostly with their use as retrofit solutions to defects. For instance, patch applications to protect crazing or small cracks (up to about 3 mm deep) can produce differential movement problems. Also, the silvery appearance of some solar coatings dulls quite quickly, and this reduces their reflective protection (a drop from 70% reflectivity to 40% in a year). The aluminium flake paints tend to form a metallic skin with different lineal coefficients of expansion and contraction from the conventional felt base covering. This can cause wrinkling and splitting at the edge of a partial repair. There can be problems with chemical incompatibility between some products.

Water ponding on the surface also causes crazing, although ponding is a surface defect. The root cause may be a defective sub-base or roof structure. The surface algae which coincide with ponding significantly reduce the solar reflectance of dressings (a drop to 20% reflectance within a year is typical). Stone chippings also help trap surface water and may reduce the frost resistance of the covering.

Inadequate bonding between the three felt layers usually used will produce hard blistering of the roof surface. Bad bonding can also be caused by poor temperature control during laying, and also by trapped moisture arising from using wet felt, by high vapour pressure inside the building or by damp driven out from the drying structure.

An empty, compressible form of blister can be caused by the pressure of vaporizing moisture trapped between layers or by poor workmanship, such as inadequate pressure during application. Materials distorted during storage may also produce defective finishes.

Projections through the roof finish are problematic because they are associated with skirtings and upstands, which are liable to vertical moisture movements and capillary effects. These sharp changes in direction localize the movement-induced stresses.

Pipes projecting through flat roofs cause many problems. Sensible dressing of the felt to the pipe may still allow water in through the gap between them. Flashings cut into the vertical surface of the pipe above the termination of the skirting may not produce an adequate seal.

## Insulation-related defects

Insulation-related defects associated with cold roof details were legendary. The energy crisis of the 1970s led to increased pressure to insulate, in many

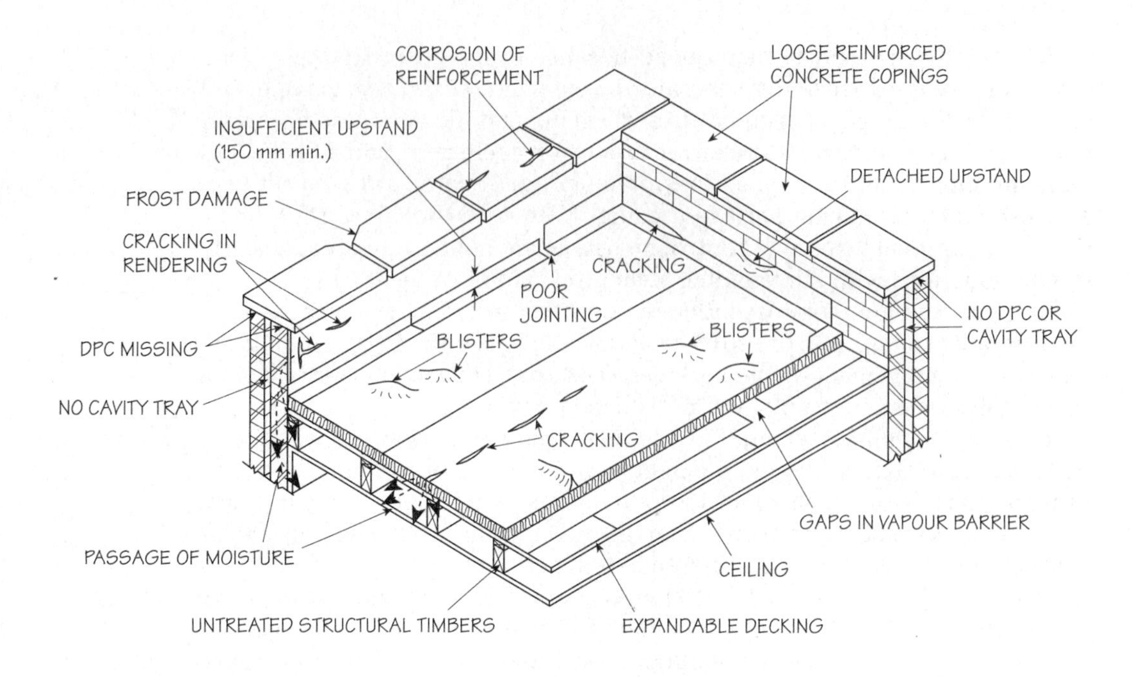


**Fig. 13.1.** Felted roofs: schematic diagram of failure mechanisms.

cases without a full appreciation of the construction methods or adequate research, and a spate of failures resulted.

The position of insulation is the distinguishing feature of the types of flat roof construction. The position of the insulation in a warm roof detail restricts heat dissipation to the substructure during hot weather. Higher temperatures occur in the covering, producing differential movement and inducing tearing stresses. This is made worse if a separating membrane is omitted.

Thermal equilibrium between the roof surface and substructure is approached only after long periods of constant temperature, and varies according to the time lag and decrement factor of the structure. Owing to variations during the day, common in the UK, steady-state conditions across roofs are rarely achieved.

## Decking and structure defects

Ponding results from an insufficient roof pitch, a flexible substrate or blocked drainage paths. Traditional flat roof structures may have shallow falls (1:100 or 1:80). With flexible substructures, a critical ponding depth must exist, beyond which the loading-induced deflection becomes self-generating.

Ridges in the covering materials are caused by unevenness or movement in the substructure and they will probably follow the pattern of joints in the substrate. Roof deck panels expand and contract during temperature cycling, and these dimensional variations in the deck can cause the covering to stretch, and then ridge up on cooling. Repetition of this increases the size of the ridge and leads to fatigue cracks at the ridge. Similar expansion ridges occur around isolated fixed points such as pipes and parapets.

The problem is most severe with warm roof details, especially if the separating membrane is omitted. The insulation layer below the waterproof covering holds the temperature of the covering high, making it more fluid.

Although most water penetration is assumed to be caused by roof finish failure, parapets are a particularly common fault. Many problems occur due to the placing, misplacing and omission of damp-proof courses in parapets. Often DPCS are omitted from the undersides of copings, and throatings may be omitted or inadequate. Parapet gutters, generally designed to serve large areas, may be defective. The risk of any overflow entering the building means that the design should be based on exceptional rainfall. Although the effectiveness of the gutter will be affected by adequate falls, it is essential that water paths are clear.

The light rendering to parapets, common with 19th century building, is liable to take up water through surface cracking. This is likely to allow water penetration into the building, where condensation on the inner leaf of cavity walls will occur. In turn, this condensation causes mould growth and low surface temperatures and forms a cold bridge.

Carrying the roof structure through the inner leaf to the cavity, or continuing it across the cavity to terminate behind brick slips, can also facilitate water travel. Stripping and replacing defective rendering may eliminate the problem of water travel, whereas cold bridging could be solved using an internal lining.

## Edge treatment

Failure at the edge of flat roofs often occurs, particularly where the discharge of water into a gutter occurs at eaves or a verge. This is generally a problem of differential movement. Joints between the head of the wall and the roof surface may be made using a timber insert or metal trim section covered with felt. The differential movements induced by this detail may cause failure cracks at right angles to the edge of the roof.

### Revision notes

- A range of defects occurring directly within the felted roof covering material, and/or the flat roof decking. Interface and differential movement/behaviour problems can be significant.
- Felted flat roofs experience a range of defects as follows.
  - Temperature-induced defects in the material, including ultraviolet degradation and oxidation. These defects may occur in combination, with the consequent embrittlement of the felt leading to the possibility of secondary defects such as water leakage (material failure).
  - Dimensional stability depends on the materials comprising the base of the felt, which have improved dramatically. Problems with dimensional instability occur at junctions and changes in direction of the supporting deck and/or upstand structure (material failure).

- Water ponding produces problems with progressive/chronic loading of the roof deck, and deterioration of the felt caused by algae build-up and/or differential movement (material failure).
- Other faults within the material include blistering, inadequate bonding; are usually related to poor workmanship (material failure).
- Problems may be created or exacerbated by high levels of insulation in warm-roof detailing, which leads to temperature build-ups in the surface material, accelerating the deformation process and maximizing the movement problems (insulation-related defects).
- Cold roof detailing produced a history of interstitial condensation and consequent structural failure in deckings in the 1960s and 1970s.
- Directly caused defects in the roof may arise from inaccuracy or inadequacy of falls to the roof, which can lead to ponding and subsequent sagging of the structure. Movement caused by changes in temperature (common with uninsulated decks) can cause tearing and compressive stressing of the felt covering (deck-related defects).

- All of these defects can occur in combination, and lead to secondary defects such as water leakage.

## Discussion topics

- Compare and contrast the categories of defect occurring in felted flat roofs. Consider which the most common and most damaging defects are likely to be.
- Consider the remedies for your identified defects. In what circumstances can the roof covering be retained?

## Further reading

BFRC *Flat Roofs: Maintenance and Repairs*, Technical Information Sheet 6.

BRE (1972) *Asphalt and Built up Felt Roofings: Durability*, Digest 144, Building Research Establishment.

BRE (1983) *Flat Roofs – Built-up Bitumen Felt – Remedying Rain Penetration*, Defect Action Sheet DAS 33, Building Research Establishment.

BRE (1983) *Flat Roofs – Built-up Bitumen Felt – Remedying Rain Penetration of Abutments and Upstands*, Defect Action Sheet DAS 34, Building Research Establishment.

BRE (1984) *Felted Cold Deck Flat Roofs – Remedying Condensation by Converting to Warm Deck*, Defect Action Sheet DAS 59, Building Research Establishment.

BRE (1986) *Flat Roof Design: Technical Options*, Digest 312, Building Research Establishment.

BRE (1987) *Flat Roof Design: Thermal Insulation*, Digest 324, Building Research Establishment.

BRE (1992) *Flat Roof Design: Waterproof Membranes*, Digest 372, Building Research Establishment.

BRE (1992) *Flat Roof Design: Waterproof Membranes*, Digest 372, Building Research Establishment.

CIRIA (1993) *Flat Roofing: Design and Good Practice*, British Flat Roofing Council and Construction Industry Research and Information Association, CIRIA Book 15.

Endean, K. (1995) *Investigating Rainwater Penetration of Modern Buildings*, Gower.

FRCAB (1988) *Built-up Flat Roofing*, Information Sheet No. 9, Flat Roofing Contractors Advisory Board.

Jenkins, M. and Saunders, G. (1995) *Bituminous Roofing Membranes – Performance in Use*, Information Paper, IP 7/95, Building Research Establishment.

Hide, W.T. (1982) *Inspection and Maintenance of Flat and Low-pitched Roofs*, BRE Information paper IP 15/82, HMSO.

Hollis, M. and Gibson, C. (1986) *Surveying Buildings*, Surveyors Publications, June.

RICS (1985) *A Practical Approach to Flat Roof Covering Problems*, Building Surveyors Guidance Notes, Royal Institution of Chartered Surveyors, Surveyors Publications.

# 13.2 Mastic asphalt flat roofs

**Learning objectives**

You should understand:

- the range of potential defects associated with mastic asphalt coverings to flat roofs;
- how to distinguish between defects originating within the waterproof covering and those associated with insulation or decking.

## Mastic asphalt failures

Any analysis of the complex mix of defects associated with flat roofs becomes a process involving detailed identification, because the faults may interact with each other or mask other faults.

Some of the more common faults with mastic asphalt coverings are summarized in Table 13.2.

## Waterproof layer

The asphalt layer rarely leaks in domestic properties, but where such defects occur they are associated with discontinuity in the decking caused by movement, or inadequate attention to changes in the direction of the covering. Skirtings and upstands are particularly vulnerable. In general, these are critical workmanship areas, and should be subject to thorough supervision and inspection.

One of the agents of deterioration-in-use of the waterproof membrane is temperature. Exposed roof surfaces are subject to the same wide temperature range as affects built-up felt roofing. Although the internal environment makes a contribution to the temperature drop through the structure, solar radiation is a major factor.

Degradation due to ultraviolet radiation hardens the asphalt so that it no longer has the necessary tensile strength to resist even minor stress. Consequently any movement of the deck or the structure can cause tensile failure of the surface. This process usually coincides with the oxidation of

*The Technology of Building Defects.* Dr John Hinks and Dr Geoff Cook.
Published in 1997 by E & FN Spon, 2–6 Boundary Row, London SE1 6HN, UK. ISBN 0 419 19770 2

the asphalt, occasionally making it difficult to determine the source of the fault.

In general, any covering without solar protection will suffer more from ultraviolet degradation than oxidation.

Asphalt becomes brittle at extremes of temperature, either because the volatile component of the material evaporates at high temperatures, or because the covering hardens at low temperatures.

Movement stresses are induced by high temperatures, the degree of stress being related almost directly to the temperature. Consequently,

**Table 13.2** Defects with mastic asphalt coverings to flat roofs

| Defect | Cause |
|---|---|
| Crazing (surface cracks) | Dressing sand not rubbed in<br>Poor solar treatment<br>Water ponding |
| Cracking (significant) | Differential movement around changes in direction and high-stress regions<br>Movement between asphalt and covering |
| Blisters | Entrapped moisture from structure or within the building |
| Ridging or cockling | Differential thermal movement between asphalt and substructure |
| Ponding | Symptomatic of:<br>• differential movement<br>• settlement<br>• poor workmanship and/or poor maintenance |
| General damage | Poor maintenance, workmanship and/or material<br>Vandalism<br>Roof traffic |
| Chemical damage | Spillage from stored containers<br>Plant and equipment |
| Loss of chippings | Severe falls<br>Wind erosion of poorly bedded chippings<br>Inadequate depth of chippings |
| Discoloured solar treatments | Surface film breaking down<br>Oxide layer forming<br>Repainting cycles too widely spaced |
| Water ingress | Cracking of surface layers<br>Inadequate or defective skirtings<br>Failure around pipes penetrating the asphalt<br>Physical or chemical damage |
| High-temperature embrittlement | Overheating asphalt during the laying process<br>Poor or inadequate supervision |
| Low-temperature embrittlement | Low ambient and structural temperature<br>Unheated building<br>Cold-roof construction |

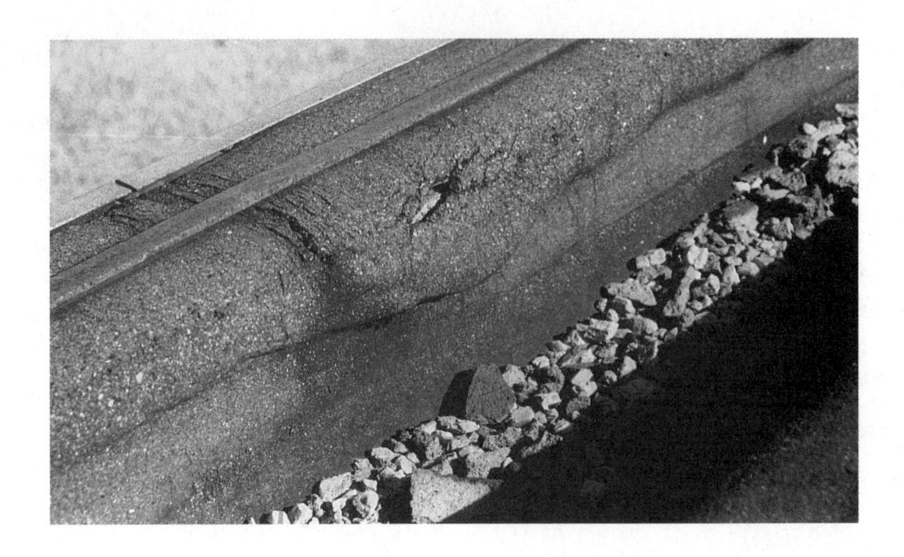

**Fig. 13.2.** Asphalt failure due to thermal distortion and movement cracking. The roof covering has failed plastically over the upstand and exposed the sub-base. (Reproduced from G. Cook and J. Hinks, *Appraising Building Defects*; published by Addison Wesley Longman, 1992.)

where deckings expand and contract, compressive stresses can be imposed on the covering. Deckings with high coefficient of linear expansion will produce the greatest degree of movement/and the greatest problems.

These temperature-related stresses act three-dimensionally, although the distances involved in the vertical direction are not often significant except for particularly vulnerable regions, such as upstands and skirtings.

The overall roof area will influence the degree of movement, and if adequate control measures are not adopted, defects such as surface crazing and cracking may occur. Cracking is concentrated around junctions with surrounding structures, at projection and around changes in the decking structure.

Water ponding on the asphalt surface can induce surface crazing. The ponding may be caused by a variety of reasons, some of them related to the decking and roof structure. The recommended deflection limit for decking is 1/325 of the span. This requirement reduces the degree of water ponding but makes it necessary to establish the nature and duration of all loading to within a relatively fine tolerance. The critical loadings are difficult to predict accurately because they are likely to be encountered under unusual circumstances and be very variable.

## Surface defects

The moulds and other growths that thrive around the fringes of water ponding areas can drastically reduce the surface reflectance of solar coatings. Stone chippings can trap surface water above the covering, but although this should increase the incidence of mould growth, the overall protective effect seems to remain high. This is due to the thermal mass of the chippings and mould growth being concentrated within them.

Moisture trapped beneath the asphalt layers produces compressible blisters, a fault caused by entrapped structural moisture or incomplete or omitted vapour barrier.

The construction of skirtings and upstands involves concave angles and these sharp changes in direction localize movement-induced stresses. Inadequately chased flashings may allow the passage of water and lead to kerbs rotting.

Pipes that project through flat roofs cause many problems, and sensible dressing of the asphalt to the pipe may still allow water in through gaps between them. Flashings cut into the vertical surface of the pipe above the termination of the skirting may not produce an adequate seal. A metal collar to maintain a seal where the asphalt shrinks away from the projection should minimize problems.

## Insulation-related defects

Asphalt roofs suffer from insulation-related defects because the trend to highly insulate the roof has produced much wider temperature ranges in use.

Warm-roof details with insulation placed immediately below the asphalt covering allow the surface temperatures to rise considerably, causing large movements that can accelerate failure.

Any projections of the substructure, such as warped edges of insulation boards, can impede asphalt movement and cause rippling.

## Decking and structure-related defects

Any unevenness or movement in the substructure can manifest itself in the covering, as shear stresses build up and cracking results. Although this generally takes longer than for felt finishes, the effects are similar.

Ridges in the covering material are rare because of the reliance on solid deckings and a range of dimensionally stable structural forms, but where the asphalt is not allowed to move independently of the deck, ripples and ridging will result.

Expansion ridges can occur around isolated fixed points such as pipes and parapets. Their causes may be traced back to defects associated with

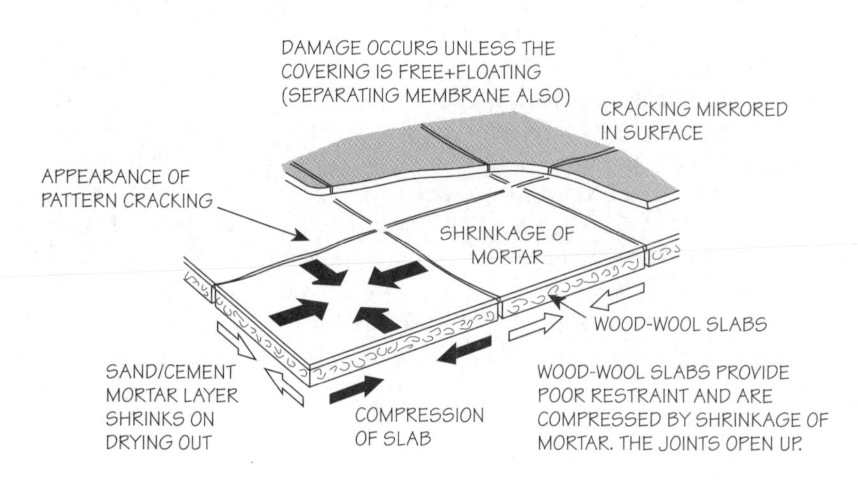


**Fig. 13.3.** Dimensional instability in roofs.

concave and convex angles formed in the mastic asphalt. These high-stress points are generally vulnerable, since they operate as the stopping and starting points of thermal and moisture movements. The high temperatures associated with warm roofs may initially assist the asphalt in enabling it to flow plastically. This may be a short-term gain since the material may reduce in thickness, thereby increasing the shear stresses operating during the next movement cycle. This can be further complicated if the separating membrane is omitted.

## Surrounding-structure-related defects

It is essential to isolate the lightweight decked asphalt roof from a surrounding heavyweight structure if failures at the junctions are to be avoided.

## Edge treatments

The stress-related defects caused by movement at the edges of asphalt roofs stem from the thermal and moisture problems in the main. However, an important area of the roof is where the water is concentrated prior to discharge.

### Revision notes

- A number of possible faults originating within the mastic asphalt, the underlying insulation or decking may appear in the covering, for example;
  - deterioration of the covering due to exposure to temperature extremes, ultraviolet radiation, oxidation of the mastic asphalt, ponding of rainwater and associated growths and differential focused movement stresses (surface defects);
  - heat- and movement-related failures due to heat build-up in insulated roof coverings, especially with warm-roof details where heat dissipation from the covering is impeded (insulation-related defects);
  - movement in the deck as a wholesale or differential release of stresses producing defects in the covering, especially at junctions and changes in direction (or rigidity of the structure); symptoms include tensile pattern cracking indicative of the structural pattern of the building/deck, or compressive rippling of the covering caused by cyclic movement (decking/structure-related defects).
- Defects may occur as a combination of these causes.

## ■ Discussion topics

- Compare and contrast the defects characteristic of felted and asphalted flat roofs. Discuss the common causes of defects in the two roof coverings.
- On the basis of the potential deficiencies of each roof covering, make the case for a choice of roofing material on the grounds of least maintenance burden.

## Further reading

Beak, J.C. and Saunders, G.K. (1991) *Mastic Asphalt for Flat Roofs: Testing for Quality Assurance*, Information Paper IP 8/91, Building Research Establishment.
BRE (1972) *Asphalt and Built-up Felt Roofings: Durability*, Digest 144, Building Research Establishment.
BRE (1992) *Flat Roof Design: Waterproof Membranes*, Digest 372, Building Research Establishment.
BSI (1970) *Code of Practice*, CP 144, British Standards Institution.
BSI (1982) *Code of Practice for Flat Roofs with Continuously Supported Coverings*, BS 622a: 1982, British Standards Institution.
CIRIA (1993) *Flat Roofing: Design and Good Practice*, British Flat Roofing Council and Construction Industry Research and Information Association, CIRIA Book 15.
Hide, W.T. (1982) *Inspection and Maintenance of Flat and Low-Pitched Roofs*, BRE Information Paper IP 15/82, HMSO.
Hollis, M. and Gibson, C. (1986) *Surveying Buildings*, Surveyors Publications, June.
Keyworth, B. (1987) Flat roof construction *Structural Survey*, **6** (2), 119.
MACEF (1994) *Trapped Water in Roofs*, Technical Information Sheet 2, Mastic Asphalt Council.

# 13.3 Metal-clad flat roofs

## Learning objectives

You should understand:

- the range of potential defects associated with metal-clad flat roofs;
- how to identify and distinguish between those defects associated with the metal covering and those related to the supporting base material and insulation.

Effective, programmed inspection is particularly important for metal roof coverings, partly because of the general belief that metal roofs have an unlimited life. Although the well-laid metal roof lasts for many years, it can suffer from design and construction defects. Most metal claddings can be damaged easily, since thin gauge sheeting is generally used.

A major factor in how long metal roof coverings last is the behaviour of the metal surface when exposed to the atmosphere – oxidation. The metal surface exposed to the atmosphere is exposed to a higher vapour pressure of oxygen than the main body of the metal, so the surface reacts with ions of oxygen to produce an oxide.

The factors which affect the durability of the oxide layer include the following.

## Stability of the oxide

This will vary, depending on the metal and the surrounding oxygen concentration – more oxygen pressure equals more stable oxide. The minor variations in atmospheric oxygen, when considered over the life of roof coverings, mean that differences on the top surface are minimal. The underside is more likely to experience oxygenation variations.

## Physical properties of the oxide

Metal oxides formed from certain metal ions occupy less space, producing porous oxide layers that allow oxidation to continue into the mass metal. Where the oxide is of similar size to, or larger than, the parent metal ion,

*The Technology of Building Defects.* Dr John Hinks and Dr Geoff Cook.
Published in 1997 by E & FN Spon, 2–6 Boundary Row, London SE1 6HN, UK. ISBN 0 419 19770 2

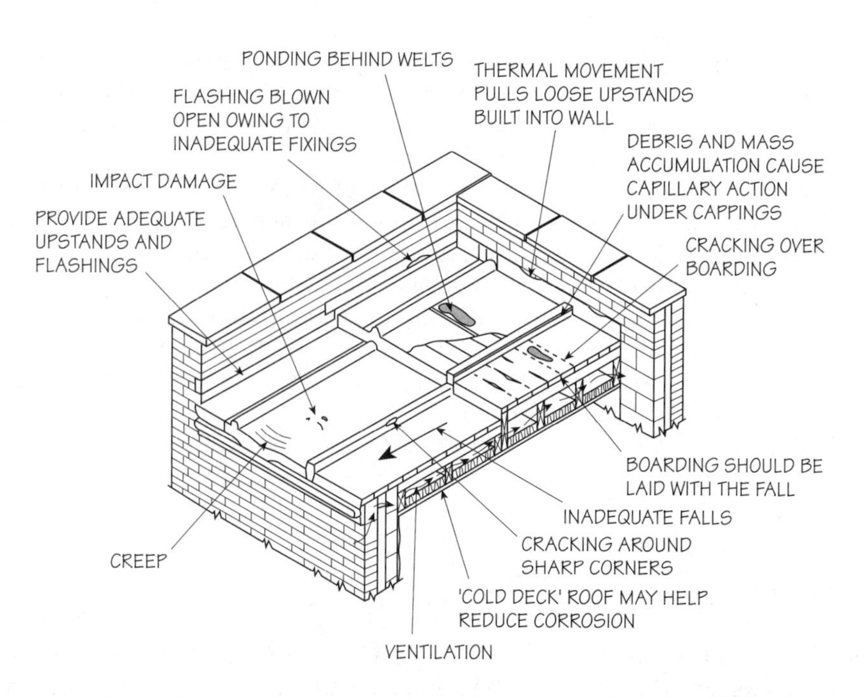


**Fig. 13.4.** Defects with continuously supported metal roofs.

then virtually impervious layers are formed (although there may be accompanying distortion and hence permeability in cases of oxides larger than the parent ion,e.g. rusting steel).

## Temperature

The higher the temperature, the higher will be the oxidation rate. So with impervious oxides, higher temperatures produce an increased depth of protection.

## Alloying

Alloying can enhance the oxidation of the parent metal, or improve its mechanical characteristics, and most roofing metals contain some alloys.

# Lead-clad roofs

## Creep

This occurs over long periods and it pulls the sheets apart, opening up joints and causing stress concentrations at junctions and changes of direction. Vertical drips are at particular risk.

## Water ingress

This major failure of the covering may be due to the following.

- Lap failure due to creep. Severe creep failure allows water to penetrate the timber decking, which causes decay.

- Leaking rolls occur when the waterproof clenching has been ineffective. Where the rolls possess arrises and irregular surfaces, then cracking can occur. Poor workmanship adds to this problem because roll dressing is an operation requiring very skilled work.
- Abutment failure allows water to penetrate the roof structure directly, short-circuiting the underlay, and spreading across the roof from the point of entry. The free-standing abutment (150 mm high recommended) allows thermal movement to occur, which when restricted by wedges can cause tearing. Where welts are provided, water can accumulate behind them, causing ponding. This water may then penetrate through minor surface defects. A similar problem can occur where the lead underlap at drips is not rebated.
- Gutters and cesspools concentrate water and so require careful detailing and construction. Debris can accumulate, causing water levels to rise above the drips and seams.

## Corrosion

The general acceptance that 'warm' roofs are superior to other construction forms has meant that where remedial treatment to existing roofs has been carried out, their features have been incorporated. Sealing of the traditional 'cold' roof ventilation path suggests that the incidence of condensation will increase at, or under, the covering and cause corrosion. Where outside temperatures fluctuate greatly, the isolation of the covering from the main building can result in water being drawn through joints owing to the reduction in pressure caused by the differences between inside and outside temperature. Once the corrosion area has become established, it will penetrate the covering. It is driven through a variety of mechanisms – low $CO_2$ concentrations, chemicals in the timber (e.g. preservatives and flame retardants) and variable oxygen levels.

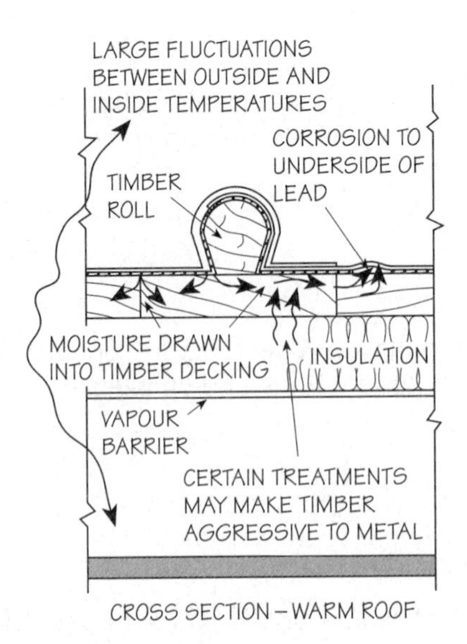


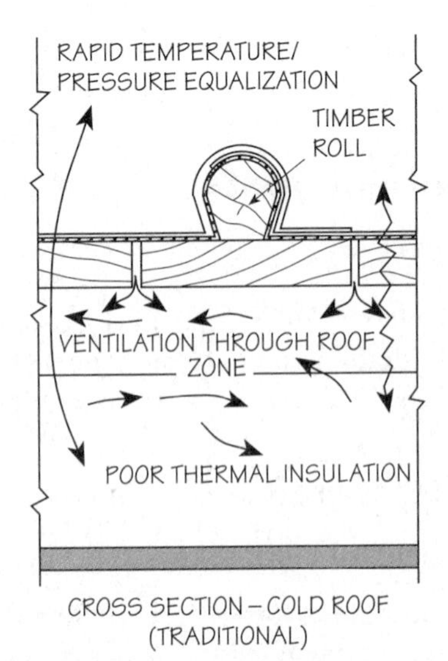


**Fig. 13.5.** Effect of roof type on potential defects.

## Copper-clad flat roofs

### Corrosion

This will normally be associated with anodic metals in contact with the copper. This will not affect the copper. Although the defect may occur some distance away from the roof, care is needed in identifying this type of defect, since concentrations in run-off water can produce corrosion. The corrosion of ferrous fixings can occur in the same way.

Certain bitumens can cause corrosion of the underside of copper sheets.

### Cracking

Work hardening around welts and seams can produce minor cracking. This may include cracking at steps and differences in the level of the substrate.

### Water ingress

This major failure of the covering may be due to a variety of causes similar to those associated with lead, for instance leaking rolls, and abutment failure due to the thermal movement caused when the material is exposed to the range of temperatures experienced by roof coverings.

## Zinc-clad flat roofs

### Water ingress

This major failure of the covering may be due to the following.

- Welted lap failure may be due to cracking caused by cold working. Where the roof fall is low, or non-existent, then water can flow back under capillary action to the substrate.
- Leaking rolls due to sharp arrises around which the sheet edges are held. Cracking at these corners may be indicative of poor workmanship, whereas sheets can become detached if adequate clipping is not provided.
- Abutment failure can occur because of thermal movement of the upstand.
- Leaking drips can occur when the weathertightness of the beaded or welted joints at the tops of the drips has failed. Where drips are provided with insufficient height (less than 50 mm, for example), capillary action can cause leakage. Ponding can occur on the feeding side of the drip, when the upstand is above the underside of the feeding sheet.
- Gutters and cesspools present similar problems to those constructed of lead or copper. The concentration of any cathodic metal in the run-off water or accumulated debris will cause corrosion.

### Corrosion

The defects associated with the adoption of a 'warm' roof are similar to those mentioned earlier for alternative coverings.

## Stainless steel flat roofs

Material failures are rare but can occur as a result of misproportioning of addictives in alloying. Excessively large sheeting may exhibit problems at fixings, including cracking. Condensation below the surface is possible, and may be a design or site fault. Other faults include the use of dissimilar metals for fixings, which can set up electrolytic corrosion cells where separating membrance have been omitted at connections.

## Revision notes

- Frequently assumed to have an unlimited maintenance free life! A number of issues require attention.
- Oxidation is a major factor determining the durability of the metal cladding. Rate and extent of oxidation depend on factors such as stability of the oxide, the physical properties of the oxide and the temperature conditions.
- Lead-clad roofs can suffer particular problems with creep and water impress due to detailing at joints and abutments. The dressing of joints is very important to prevent water impress and for entrapment.
- Copper-clad roofs can be affected by contact with certain bitumens, otherwise corrosion to the roof covering in the conventional sense tends to happen to other metals in direct or indirect (limited by water run-off) contact with the copper roof. Copper can be vulnerable to cracking at points which concentrate stresses and produce a work-hardening effect. The joints are critical, as with lead roofs, and the fault mechanisms are similar.
- Zinc-clad roofs also exhibit similar defects abutments and guttering or cesspool details where changes in profile and/or connections create locations potentially subject to stresses and/or leakage.
- Stainless-steel-clad roofs may exhibit deficiencies in the material itself; or because of design/construction faults, a range of splitting and corrosion problems may arise (corrosion specifically where fixings have been made without isolating dissimilar fixing metals).

## ■ Discussion topic

- Compare and contrast the symptoms and mechanisms of failure in lead, copper, stainless steel and zinc clad flat roofs and comment on each material's nature of deficiency in use.

**Table 13.3** Metal roofs: symptoms and causes of defects

| Defect | Cause |
|---|---|
| Lead | |
| Splits and cracks | Insufficient thickness<br>Underlay softening, causing lead to stick, restricting thermal movement<br>Incorrect joint spacing |
| Dents and cuts | Physical damage |
| Edges lifting | Wind action |
| Surface marking, loss of metal | Chemical attack (electrolytic corrosion) from external sources |
| Sugaring to underside of metal | Corrosion from internal sources[a] |
| White streaks (corrosion) | Concentrated water flow |
| Movement of lead | Creep of the lead |
| Copper | |
| Patina or verdigris | Oxidation; the amount of atmospheric pollution affects rate of patination |
| Corrosion of fixings, external | Adjacent to cathodic metal, e.g. aluminium |
| Corrosion from internal sources | Condensation, which may cause failure of organic substrate, e.g. timber<br>Chemical incompatibility with elements in bitumen felt |
| Cracking around seams and rolls | Over-working causing localized hardening |
| Water ingress | Single cross-welts provided on pitch less than 45°<br>Corrosion from adjacent cathodic metal<br>Movement of sheet edges due to wind action |
| Dents and cuts | Physical damage |
| Edges lifting | Wind action |
| Zinc | |
| Surface corrosion and pitting | Atmospheric pollution<br>Water draining from adjacent cathodic metal roof or fittings |
| Internal surface corrosion | Condensation<br>Corrosion from acidic timber decking |
| Water ingress | Leaking welted lap<br>Insufficient clipping<br>Splits and cracks<br>Capillary action under cappings<br>Nailing through cappings and sheets |
| Dents and cuts | Physical damage |
| Edges lifting | Wind action |

[a] The risk exists of water being drawn in through the joints and rolls from outside, particularly where a warm-deck roof construction has been adopted.

## Further reading

Ashurst, J. and Ashurst, N. (1988) *Metals*, Practical Building Conservation Series, English Heritage Technical Book 4.
Building Supplement (1987), *Architects Question Warm Deck*, 29 May.
CIRIA (1993) *Flat Roofing: Design and Good Practice*, British Flat Roofing Council and Construction Industry Research and Information Association CIRIA Book 15.
LSA (1992) *Lead sheet roofing and cladding*, in *Lead Sheet Manual*, Lead Sheet Association, Vol. 2.
LSA (1993) *Underside Corrosion*, Update 2, Lead Sheet Association.
RICS (1985) Flat Roof Covering Problems, Building Surveyors Guidance Notes.
Taylor, G.D. (1991) Materials of Construction, 2nd edn, Construction Press.
Whatford, M. (1985) *Fully-Supported Stainless Steel Roof Coverings – Typical Defects that can Occur*, Building Surveyors Technical Paper, Royal Institution of Chartered Surveyors, Surveyors Publications.

# 13.4 Pitched roofs

## Learning objectives

You should understand:

- the range of defects occurring in the structure of the pitched roof;
- its relationship with defects in the covering;
- its relationship with the remainder of the building structure.

Modern pitched roofs generally consist of a timber frame covered with an impermeable barrier that is composed of small units such as tiles or slates – though other traditional coverings such as thatch and cedar shingles can also be found in use.

Trussed rafter roofs require careful storage and handling if they are to perform satisfactorily. Poor storage leads to a high moisture content that will not reduce rapidly in poorly ventilated and highly insulated roofs. Older

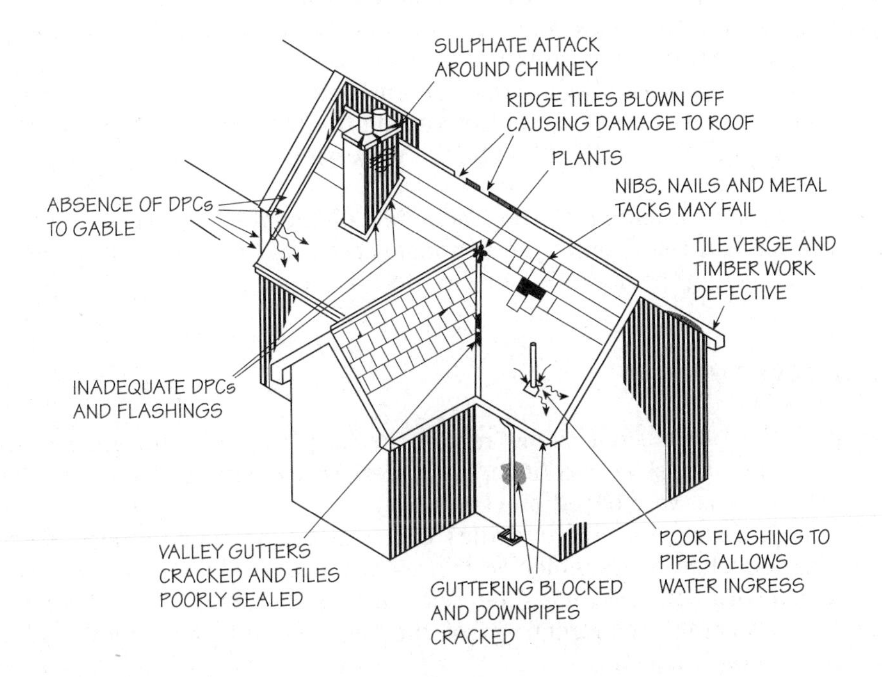


**Fig. 13.6.** Location and nature of pitched roof defects.

*The Technology of Building Defects.* Dr John Hinks and Dr Geoff Cook.
Published in 1997 by E & FN Spon, 2–6 Boundary Row, London SE1 6HN, UK. ISBN 0 419 19770 2

roofs are likely to have been framed in situ and rely on the quality of the workmanship to perform satisfactorily.

## Clay tiles

Clay tiles are still made but their expense has led to old tiles being reused. It is, however, essential that badly weathered or defective tiles are rejected. Underfired tiles can delaminate because of frost, but this is difficult to assess (although evidence of satisfactory performance is a good guide).

In general the single-lap tiles are suitable for steep pitches of more than 40°. Where the pitch is shallow, water may enter through the laps, owing to an increased angle of creep.

The fixing of ties to tiling battens using nails can fail because of continual condensation on the underside of the roof covering.

The nailing positions are always covered by two other tiles and so replacement of tiles requires support from the tiles (usually lower) next to them. In many cases defects with tiles are difficult to repair without causing further problems.

Reduced ventilation between the tile and the underlay can cause this fault, which in turn can lead to increased susceptibility to frost attack. The problem tends to increase as the pitch reduces. Adequate lapping and support of the underlay felt, particularly at the eaves, is essential to stop water getting in.

Interlocking tiles and tiles having a single overlap must have a felt underlay capable of handling a large amount of water (the amount of water handled is greatest at the lower section of the roof). Correcting the ridge tile pointing can sometimes cause more problems than it is worth, particularly since the amount of water admitted is low.

Tiles may have nibs to locate them on the battens but these can fail with time. The nibs can be inspected from inside the roof space but, where the roof has been lined with felt, tapping the underside will show any deterioration of the nibs.

Pargeting, the pointing of the underside of the tiles to prevent water penetration, can also complicate any inspection. But it is likely that where there is pargeting the condition of the roof will be suspect.

## Concrete tiles

Many of the points concerning repairs for clay tiles can be applied to concrete tiles too, since although the materials are different, many of the concrete designs were based on clay models.

The high density of concrete tiles can lead to the roof timbers being overstressed where roofs which were previously covered with clay tiles are re-covered with concrete tiles. The risk can be structurally determined, so that any necessary reinforcement of the roof can be established under reduced imposed loads.

Acidic atmospheric pollution will slowly etch cement away from the upper surface of the tile, so that it accumulates in gutters and downpipes.

## Slates

Slating battens are difficult to reach, so replacing slates can be difficult. Lead tacks are used, although these can fail owing to wind action and age. Slates become brittle with age so it becomes more difficult to replace slates without breaking the slates next to them. Care and attention are needed, particularly where the surrounding slates are providing support for the repair. Severely defective roofs may need total re-roofing.

Lead soakers at junctions and abutments are essential but although they are workable and durable, they suffer similar defects to metal roof coverings.

## Insulation-related defects

Insulation in pitched roofs at ceiling level can increase the incidence of condensation, which can be reduced by adequate roof ventilation.

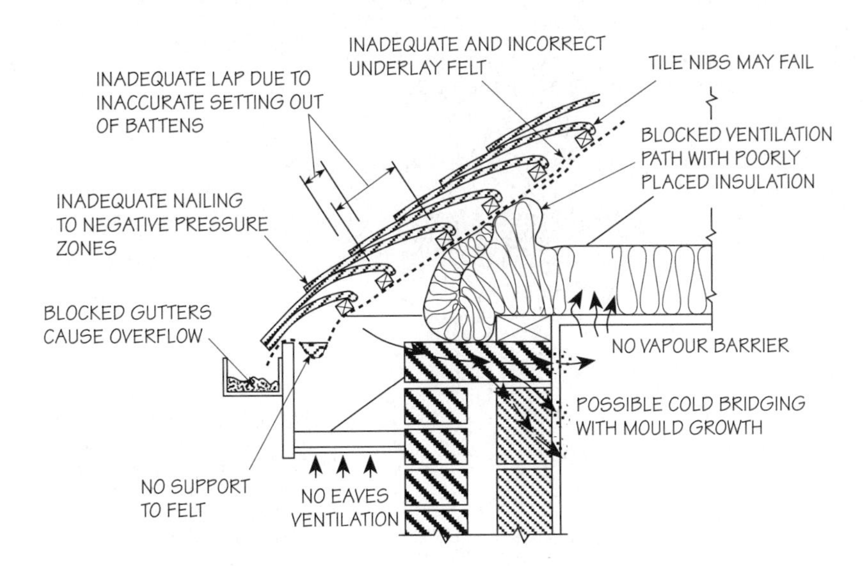


**Fig. 13.7.** Potential defects at eaves of pitched roofs.

Condensation creates conditions suitable for fungal attack and corrosion of metal fixings and so weakens the roof timbers. Where ventilation has been provided, this can become blocked by insulation or stored items. Masonry carried up to the underside of the roofing felt causes the same effect.

## Sagging and hogging in pitched roof structures

The rigidity through triangulation in traditional roofs with purlins and ridge plates meant that a roof in good condition would be self-restraining. Rotting in timber roof frames at connections or bearing points can lead to unbalanced forces being imposed on the supporting walls, as the roof transfers loading and sags. In essence, the well-braced and triangulated structure can revert to a series of two-dimensional unbraced components which are unstable. It is normal for roof frames (and floors) to be structurally supported on the inner leaf of cavity walls, and it would be usual for distortion and cracking to appear in the inner leaf. The eaves area is especially vulnerable because of the low resistance to side forces coupled with unusual self-weight.

Racking can also occur as the failure (or omission) of bracing allows the roof structure to distort, or lozenge. A 40 mm distortion from vertical may be all that is needed in an unbraced (or effectively unbraced) timber trussed roof for structural integrity to be compromised. In such an instance the roof may transfer loading to the gable, usually a relatively weak and slender wall. The result can be bending in the gable as it attempts to carry side forces from the racking roof for which it was not designed.

Many gable walls were strengthened by use of a chimney structure, and may be more able to carry the load. Note that where the lower part of the chimney has been removed there may be bending in the upper part of the gable which is *causing* a problems in roof stability rather than *reacting* to it.

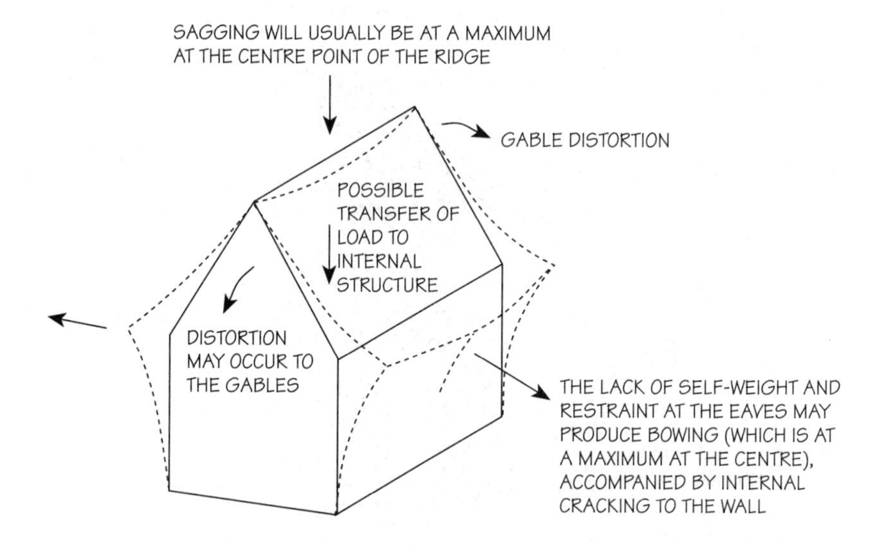


**Fig. 13.8.** Tent-like sagging in pitched roof structure.

Sagging in roofs tends to be brought on by failure of the tensile stiffening in roofs, either the ceiling joists or ties in braced trusses. Sagging may also be simply due to overloading, via a mechanism similar to bulging in walls due to floor deflection and load transfer.

Externally, the symptoms will be a tent-like bowing of the ridge coupled with outward deflection of the eaves, particularly near the mid-point. The response of the gables will depend on the strength and stiffness of the wall, which will be carrying increased load including side forces. Symptoms similar to those caused by racking may occur in the gable(s).

Internally, the eaves walling may exhibit cracking together with ceiling deflection and/or damage. Loads may be transferred to internal walls, producing further deflection and other symptoms such as problems with closing doors.

**Fig. 13.9.** General sagging and distortion of a roof structure. This may be due to overloading or failure of the structure. (K. Bright.)

Roof structures may undergo long-term creep. This may produce (relatively) minor distortions which result in internal non-loadbearing partitions carrying transferred loads. In Victorian buildings, for example, the internal timber walls were designed as loadbearing. Interference with or removal of these walls can therefore be a cause of subsequent roof sagging.

Insect or fungal attack of the roof timbers will seriously affect their structural performance. Eradication and timber replacement may require removing the roof covering to provide access and reduce the stresses within the structure.

These stresses may have been sufficient to push out the existing walls, which may require rebuilding as part of an overall renovation programme of the roof.

## Problems with roof timbers

The difficulty in replacing structural timbers is due to the need to relieve loading during the operation. In certain older buildings there may also be a general reluctance to remove or replace parts of visible elements such as roof trusses which may have historic value.

Any localized stiffening of the timber may cause movement stresses, and the change in thermal conductivity and/or vapour permeability could lead to condensation inside the timber at the interface between treated and untreated parts, possibly initiating new decay.

## Other problem areas

Rainwater causes a high proportion of the outbreaks of dry rot. Defective valley gutters (9%) and other gutters (16%) are significantly frequent. Other roof problems (11%) and parapet faults (3%) have been recorded. Out of all the cases of dry rot in building timbers, 23% were associated with pitched roofs and 5% with flat roofs.

The intersection points on complex room shapes such as dormers can be areas where significant problems can occur, mainly because of the weathering and flashings around the openings. Gutters can also become blocked, causing water ingress to the roof space.

Punctured valley or parapet gutters and blocked downpipes are focus points for water entering the roof or building. The correct depth and falls should be provided and regular cleaning is essential. Timber support systems can fail, since they are as vulnerable as the main roof timbers.

Projecting features, and their associated joint with the roof covering, are also vulnerable areas. The weatherproofing at the back of chimneys is a common failure zone. The area is difficult to inspect, and may have been designed as a flashing, rather than a gutter.

Additional problems can arise at the junction between a tiled roof and a wall. These joints must be flexible, and where cement render is used, failure will occur unless there is additional protection from metal soakers.

**Fig. 13.10.** Distortion of roof coverings over a party wall caused by differential movement of the main roof areas. (K. Bright).

**Table 13.4** Pitched roofs: general defects

| Defect | Cause | Remedy |
|---|---|---|
| Roof units loose and/ or slipped | Corrosion of fixings | Replace fixing with durable alternative |
| | Lack of fixings | Provide fixings, using Al, Cu or stainless steel, as BS 5534 |
| | Severe wind loading | Replace, check the design with regard to CP3<br>This can also cause wear around the nail hole, causing failure |
| | Deterioration of the nibs | Replace individual tiles<br>Where nibs are not integral, they may be replaced |
| | Deterioration of tile battens | Remedy condition for fungal attack<br>Replace defective battens |
| Sagging of roof | Distortion of roof timbers, due to long-term movement | Remedial work will only be necessary when structurally unstable<br>Take long-term measurements<br>Provide permanent propping |
| | Roof timbers deteriorating (fungal and/or insect attack) | Remedy cause of attack<br>Proceed as for defective timber<br>Minor cases may be treated with timber patching |
| | Corrosion of metal fixings | Remake joints with non-ferrous fixings<br>Patching of gussets may be required |
| | New roof covering heavier than original | Remove and replace with lighter covering |
| Roof spread causing cracking | Shrinkage of wall plate | Make good around wall plate |
| Roof spread causing cracking and outward movement of wall | Underdesign of roof timbers | Provide additional structural members in roof |
| | New roof covering heavier than original | If severe replace by lighter covering |
| Deterioration of timber (fungal/insect attack) | Remedy timber attack, replacing timber where attack is severe<br>Metal rods, plates and adjusting nuts can be used to pull walls back to the vertical | |
| Lamination and spalling of clay tiles | Frost action | Short-term replacement is the precursor to total roof replacement |
| | Low roof pitch | Run-off times are extended<br>Consider replacement |
| Lamination of natural slates | Atmospheric pollution | Most indigenous slates are resistant<br>Check country of origin<br>Replace separate units although the whole roof may eventually fail |
| | Frost | Rare, slates of poor quality<br>Replace |
| Deterioration of cedar shingles | Fungal attack, due to being located in an area of high rainfall | Remove shingles and replace with preservative-treated type |

Pipes projecting through the covering require care in construction if water is not to penetrate through.

## Revision notes

- Usually the structure of the roof is timber. Modern pitched roofs based on trussed construction are highly dependent on correct bracing, and have a low tolerance of distortions in their verticality.
- The interface between the roof structure and the coverings such as tiles or slates can produce problems with water leakage, particularly at edge details and changes in roof profile. Problems may arise with fixing of roof coverings also.
- Defects associated with insulation may include condensation in the roof space, which may be made more problematic where ventilation has been restricted.
- Roofs and walls should be connected with strapping. Racking of roofs may lead to distortion in walling, especially slender gable walls (and vice versa). The result may be lozenging of the structure as it moves off verticality.
- Roof structures may also exhibit sagging, which can result in the transfer of loads to the remainder of the structure. Long-term creep of the frame can also be problematic.

## ■ Discussion topics

- Compare the causal factors affecting the performance of flat roofs and pitched roofs, and analyse the mechanisms by which failures tend to occur.
- Discuss the range of defects in the main structure of a building which will produce symptoms in the roof structure/covering, and how to distinguish between them.

## Further reading

BRE (1982) *Pitched Roofs – Soaking Felt Underlay – Drainage from Roof*, Defect Action Sheet DAS 9, Building Research Establishment.

BRE (1982) *Pitched Roofs – Thermal Insulation Near the Eaves*, Defect Action Sheet DAS 4, Building Research Establishment.

BRE (1983) *External and Separating Walls – Lateral Restraint at Pitched Roof Level: Specification*, Defect Action Sheet DAS 27, Building Research Establishment.

BRE (1983) *External and Separating Walls – Lateral Restraint at Pitched Roof Level: Installation*, Defect Action Sheet DAS 28, Building Research Establishment.

BRE (1985) *Slate or Tiled Pitched Roofs – Ventilation to Outside Air*, Defect Action Sheet DAS 1, Building Research Establishment.

BRE (1985) *Slate or Tiled Pitched Roofs – Restricting the Entry of Water Vapour from the House*, Defect Action Sheet DAS 3, Building Research Establishment.
BRE (1986) *Dual-Pitched Roofs – Trussed Rafters – Bracing and Binders: Specification*, Defect Action Sheet DAS 83, Building Research Establishment.
BRE (1986) *Dual-Pitched Roofs – Trussed Rafters – Bracing and Binders: Installation*, Defect Action Sheet DAS 84, Building Research Establishment.
BRE (1991) *Bracing Trussed Rafter Roofs*, Good Building Guide 8, Building Research Establishment.
BRE (1993) *Supplementary Guidance for Assessment of Timber-Framed Houses, Part 2. Interpretation*, Good Building Guide 12, Building Research Establishment.
TRADA (1993) *Principles of Pitched Roof Construction*, Wood Information Sheet 10, Section 1.

# 13.5 Roofing slates, clay roofing tiles and concrete roofing tiles

## Learning objectives

You should be able to:

- explain the causes of nail and nail hole failure of slates;
- describe the causes of delamination of slates and clay roofing tiles;
- discuss the relationship between porosity and frost resistance of clay roofing tiles;
- explain the causes of chemical deterioration of clay roofing tiles;
- describe the causes of deterioration of concrete roofing tiles;
- compare the causes of deterioration of slates, clay roof tiles and concrete roof tiles.

## Problems with slates

Slate is a metamorphic stone formed from clay or shale-like material by heat and/or pressure. This rock has clearly defined cleavage planes and these are used to split the rock into thin sheets for use as roof coverings. The cleavage planes are also a region where delamination can occur. Welsh slates of 3–5 mm thickness have a variety of low-chroma colours including grey, blue and green. Cornish slates are generally green-grey. Thicker sections of slate are used for window cills, vertical claddings and pavings. The tensile strength of slate, although good when compared to the other stones, is significantly lower than the compressive strength. A general weakness in the cleavage plane could be detected by the traditional method of assessing the quality of slates. This suggests that they should 'ring true when struck'.

The durability of the material is due to good chemical resistance, very low water absorption and virtual impermeability. This quality of durability endorses the general advice that where slates have slipped off roofs it is the nails that should be replaced, not the slates. Where this has occurred over many years even slates may start to deteriorate. Isolated nail failure in

*The Technology of Building Defects.* Dr John Hinks and Dr Geoff Cook.
Published in 1997 by E & FN Spon, 2–6 Boundary Row, London SE1 6HN, UK. ISBN 0 419 19770 2

slates or tiles can be refixed with metal tingles. This may indicate a general ageing of the nail fixings across the roof. Where nail holes are not intact the nails may have worn the slate away owing to slight movement caused by wind action.

The dimensional stability is excellent since moisture movement is assumed to be zero and the coefficient of linear expansion is around $8 \times 10^6$ per °C.

Delamination can occur with slates, although this is more common with inferior quality slates. These may contain calcium carbonate, which under weathering converts to soluble calcium sulphate.

Failure of slates in use can be due to deterioration of the fixing nails. This may be due to corrosion of steel nails or wear induced by wind action.

## Problems with clay roofing tiles

Clay roofing tiles have similar durability characteristics to slates. Water absorption is an important durability factor when considering roofing materials. This influences the ability of the material to shed water rapidly and, since frosts can follow rain or snow thaw, to resist the deteriorating effects of frost. Clay tiles may have a water absorption of 10.5% of the dry weight. The frost resistance of porous materials is complex and does not always follow the general rule that low porosity provides good frost resistance. Hand-made types can be more porous than some types of machine-made tiles and may have a higher resistance to frost attack. This is evidenced by some types of machine-made clay tiles failing by delamination. The lower the roof pitch the slower the water run-off rate and therefore the greater the risk of water absorption by the tiles. The traditional plain tiled roof would have the benefit of a steep pitch, perhaps as much as 45°, and a slight camber on the tile which afforded capillary separation between tiles.

Where tiles are inadequately fired they may contain soluble salts. These can be leached out during weathering, increasing the porosity of the tile and

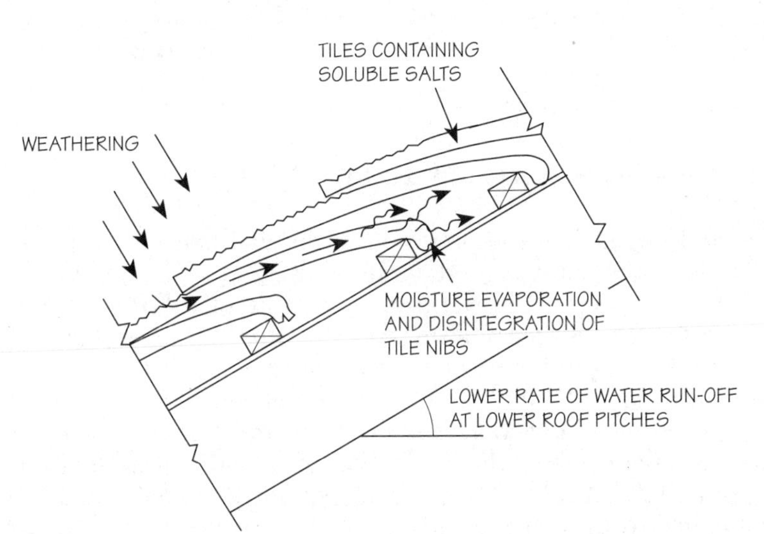


**Fig. 13.11.** Disintegration of tile nibs.

lowering the frost resistance. This leaching out can occur in that part of the tile which is protected by overlapping tiles. Although normally a deficiency of relatively few tiles in a batch, this can cause the disintegration of tile nibs.

The coefficient of linear expansion is slightly lower than that of slate, being around $5–6 \times 10^6$ per °C. The movement effects are therefore confined to wind action.

## Problems with concrete roofing tiles

Concrete tiles are widely used. These are precast and contain small-sized aggregate. Although the density of concrete, around 2400 kg/m$^3$, is slightly lower then that of slate, around 2590 kg/m$^3$, the increased thickness of concrete tiles means that where slate roofs are re-tiled there is an increased load on the roof timbers. This may exceed the design load and result in distortion or failure of the roof timbers.

Concrete tiles are affected by the same range of general defects which can affect concrete. Since they are precast in a factory environment they should not experience defects associated with site mixing. Sulphate attack is possible around leaking flues or poorly designed flue terminations. Carbonation can be considered a minor problem since roof tiles are unreinforced. Cracking can occur in concrete tiles since they are relatively thin and brittle. This can be critical where the crack occurs on or across the side interlock in single-lap tiles. Poor storage and damage during fixing where tiles span between battens can cause cracking. General maintenance work involving pedestrian access can be sufficient to crack roofing tiles.

Surface erosion of the tiles, particularly in the early life of the tiles, can block valley and eaves guttering when aggregated with dust from the general environment. The corresponding change in colour to generally darker shades may not affect the performance of the tile, although it can complicate colour matching with any new extensions to the roof.

Although the maximum aggregate size should not be greater than 4.75 mm, relatively large aggregate particles serve to concentrate frost-induced defects. Localized spalling of the tile can occur which exposes the aggregate particle within the tile.

## General defects with slates and tiles

The general increase in the provision of thermal insulation for roofs may have implications for the deterioration of tiled roof coverings. Where insulation is placed on the warm or inner side of the roof then the roof covering is likely to remain at lower temperatures for longer periods of time. Frost action is therefore more likely to occur, which could have implications for the long-term deterioration of tiled roof coverings.

Roof pitches have generally reduced over the last three decades. This appears to have occurred for economic, technical and aesthetic reasons. Whilst some roof tiles have been developed for this application, e.g. interlocking concrete tiles, the origins of other types of roof tiling are more

**Fig. 13.12.** General slippage of roof tiles due to nib and nail failure. (D. McGlynn.)

directly aligned with steeper pitches. The rate of water run-off from tiles is related to the roof pitch and for lower pitches this may mean that tiled roof coverings are exposed to a greater amount of weathering. This may have implications for the long-term durability of some tiled roofing systems.

Tiles and slates can be blown off roofs when not nailed. This can also occur where there is a reliance on the tile nib over the batten to retain the tile. The nibs on clay and concrete tiles can crack off where the clay is brittle owing to ageing. A small tile or slate lap will mean that there is less mass to resist the effects of wind action. Since laps are generally smaller on steeply pitched roofs, and more of the roof surface is likely to be in regions of negative pressure under wind action, this is more likely to occur on steeply pitched roofs.

## Revision notes

- Slate, clay tiles and concrete tiles are less able to resist tensile stress than compressive.
- Slates have cleavage planes which can, under frost or chemical attack, cause delamination. Those slates which contain calcium carbonate can weather to produce soluble calcium sulphate and this may contribute to delamination.
- Nail fixings can either corrode or abrade through slates under the action of the wind. Metal tingles can indicate the former as well as indicating the scale of the deterioration.
- Clay tiles may absorb 10% of their dry weight, which can adversely affect their frost resistance. However the link between absorption and frost resistance is not straightforward. Large aggregate particles in concrete tiles can concentrate frost-induced deterioration.

- Plain clay tiles are normally used on steeply pitched roofs. Where they are used on shallow pitches they may be less exposed to wind action but more exposed to water absorption. Frost-induced deterioration may be more likely.
- Poorly fired clay tiles may contain soluble salts; these can increase the porosity of the tile when weathered.
- Concrete tiles are heavier than similar-sized clay tiles or slates; this can overload the supporting roof structure.
- Concrete tiles can be subject to the normal range of defects associated with concrete. Brittle failure and cracking is a particular problem.
- Where insulation is placed on the warm or inner side of the roof then the roof covering is likely to remain at lower temperatures for longer periods of time. Frost action is therefore more likely to occur.

## ■ Discussion topics

- Compare the deterioration mechanisms of fixings for slates, clay tiles and concrete tiles.
- Discuss the implications of the deterioration of tiled roof coverings for the deterioration of the roof structure.
- Describe the chemically induced defects which can occur in clay and concrete roof tiles.
- Critically appraise the implications of adopting low roof pitches and high thermal insulation in roofs for the deterioration of tiled roof coverings.

## Further reading

BRE (1964) *Design and Appearance – Part 2*, Digest 46, Building Research Establishment, HMSO.

BRE (1983) *The Selection of Natural Building Stone*, Digest 269, Building Research Establishment, HMSO.

BRE (1984) *Wind Loads on Canopy Roofs*, Digest 284, Building Research Establishment, HMSO.

BRE (1994) *Damage to Roofs from Aircraft Wake Vortices*, Digest 391, Building Research Establishment, HMSO.

BRE (1991) *The Weathering of Natural Building Stone*, Building Report 62, Building Research Establishment, HMSO.

Cook, G.K. and Hinks, A.J. (1992) *Appraising Building Defects: Perspectives on Stability and Hygrothermal Performance*, Longman Scientific & Technical, London.

King, H. and Everett, A. (1971) *Components and Finishes – Mitchells Building Construction*, Longman Scientific & Technical, London.

PSA (1989) *Defects in Buildings*, HMSO.

Richardson, B.A. (1991) *Defects and Deterioration in Buildings*, E. & F.N. Spon, London.

Taylor, G.D. (1991) *Construction Materials*, Longman Scientific & Technical, London.

CHAPTER 14

# Radon in buildings

## Learning objectives

- A naturally occurring radiation problem which enters the building via permeable floors and walls.
- You should understand the mechanism of entry and potential problems with the presence of radon in buildings.

Radon is a feature of the background radiation levels. It is a dense gas released as uranium-238 present in the ground and traditional building materials decomposes. Radon is radioactive, and a risk to health from it and its by-products has long been established from mining incidents. Across the majority of the UK the background radiation level is relatively low.

## Entry into the building

The gas passes through any permeable layers in the structure from the ground and into the internal spaces, particularly from the floor. The dissipation of the gas within the internal spaces will of course depend on the ventilation rate, and where this is low there is the possibility of build-up and increased exposure of the occupants.

## Radioactivity mechanism

Radon embedded in materials such as concrete and bricks is a primary source of radioactivity. As it decays it produces penetrating gamma rays. This radioactivity will pass through building materials and the occupants are directly exposed.

The decaying radon gas releases polonium, another carcinogen. The mechanism involves dust carrying the polonium being breathed in and settling in the upper lungs. Alpha rays released from the collecting polonium damage the skin and produce cancer. It is estimated that in total 600 lung cancer deaths per year can be attributed to radon via polonium. This is a higher risk than that from asbestos, and the levels of radiation involved are many times greater than those allowed for radioactive discharge.

## Construction factors

Factors which will increase the risk are floors to houses which have no barrier to the radon gas, exacerbated by any features of design or use that

*The Technology of Building Defects.* Dr John Hinks and Dr Geoff Cook.
Published in 1997 by E & FN Spon, 2–6 Boundary Row, London SE1 6HN, UK. ISBN 0 419 19770 2

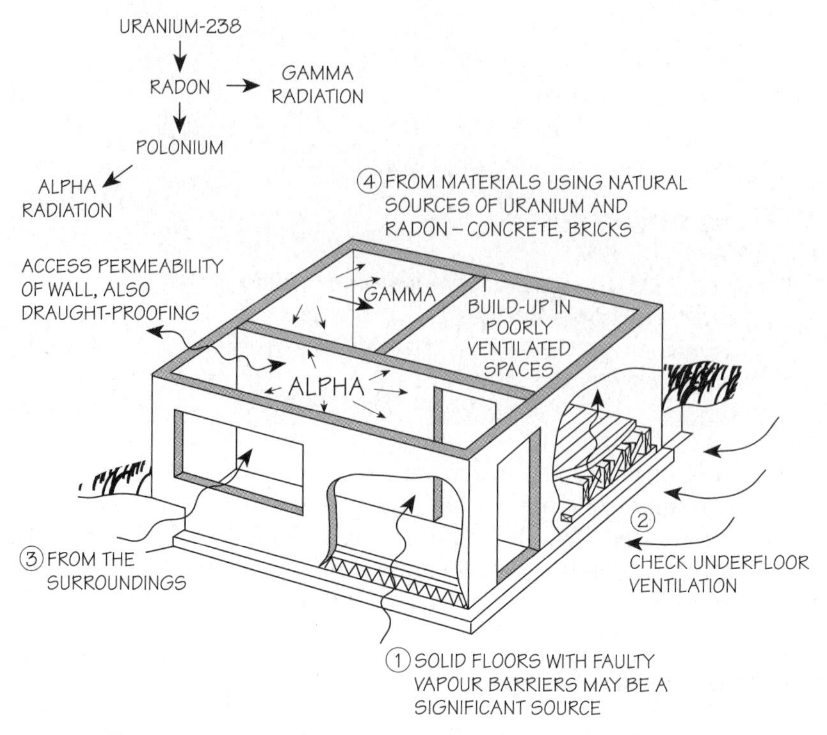


**Fig. 14.1.** Sources of radon.

minimize the ventilation rate. Common examples include insulated walls, which have a reduced permeability, and also draught-proofed openings or double glazing. Dense concrete floors with a low vapour permeability or floors with effective vapour barriers should be of minimal agency to radon transmission, but this will be affected by any cracks and openings in the floor slab. Consequently, it can become the biggest single source. The external background levels in the air will also mean that some gas will be introduced into houses which themselves are not contributory sources.

The materials used in the construction of traditional building products may contain uranium and decay to release radon and radioactivity directly into the house. For example, phosphogypsum used for the widespread production of building boards has been identified as an acutely polluted material.

Obviously the nature of the ground under and surrounding the building will be important. High uranium concentrations occur in the Highlands and Cornwall, and the presence of significant concentrations of radon and its by-products is likely.

## Revision notes

- A background radiation phenomenon, and therefore a health problem.
- Radon gas is released as uranium-238 decomposes.
- Radon enters buildings via permeability in the envelope, particularly floors.
- Sources include the ground, the construction materials and background sources in the air.
- Direct injury via gamma radiation, and alpha rays released from polonium, which is a decay product of radon.
- Buildings with poor ventilation tend to allow high concentrations of radon to build up.

## ■ Discussion topics

- Describe the mechanisms of entry and concentration of radon, and how this may be dealt with in existing properties.

## Further reading

BRE (1991) *Radon: Guidance on Protective Measures for New Dwellings*, Report 211, Building Research Establishment.

BRE (1992) *Radon Samps: Guidance on Protective Measures for Existing Dwellings*, Report 227, Building Research Establishment.

DOE (1993) *Householders Guide to Radon*, 3rd edn, HMSO.

Green, B.M.R., Lomas, P.R. and O'Riordan, M.C. (1992) *Radon in Dwellings in England*, Report 254, National Radiological Protection Board.

NRPB (1990) *Board Statement on Radon in Homes*, Documents of the NRPB Vol. 1/1 National Radiological Protection Board.

NRPB (1990) *Radon Affected Areas: Cornwall and Devon*, Documents of the NRPB Vol. 1/4 National Radiological Protection Board.

Scivyer, G.R. (1994) *Surveying Buildings with High Indoor Radiation Levels – BRE Guide to Radon Remedial Measures in Existing Dwellings*, Report 250, Building Research Establishment.

CHAPTER 15

# Distortion in buildings

## Learning objectives

You should understand:

- the principal causes of distortion in the structure of buildings;
- the likely locations of distortion;
- the effect of the characteristic features of the building on the manifestation of defects – including connectivity, mass, restraint and stress concentration phenomena.

Two principal causes of distortion in the structure of buildings are overloading and internally developed instabilities. Either may be caused by inadequate attention during design or construction of the building. Also, internally induced stresses in the building materials arising from moisture and thermal movement can cause deformations as the elements elongate or shorten.

Creep-related or differential movements in the structure can arise from the self-weight of the elements, imposed loads or an imbalance with the supporting soil. The failure symptoms represent the relief of otherwise unresolved forces. The relieving effect of cracking or deflection may be sufficient to leave the fault stable. Alternatively, residual stresses will produce progressive failure.

## Location of stresses

It is important to identify the source of stresses which can produce cracking, bulging or differential movement at the weakest point in the local structure. The exact nature of failure will depend on the constructional form and the material characteristics. Stress concentrations will usually locally overload the structural materials. The connections between different materials will frequently become the focus for movement-related stresses. In such circumstances joint failure is likely. Hence string courses in clad elevations will tend to concentrate the movement in the frame. Shear failures may occur at the junctions between a frame and the cladding. In the case of rigid brick cladding enclosed in a concrete frame, the contrasting expansion and shrinkage produce tensile failure in the frame or bulging of enclosed brickwork as its free movement is restrained (see elsewhere, this work). Detachment cracking and differential movement will be concentrated at the junction between heavy or stiff components and the weaker or flexible

*The Technology of Building Defects.* Dr John Hinks and Dr Geoff Cook.
Published in 1997 by E & FN Spon, 2–6 Boundary Row, London SE1 6HN, UK. ISBN 0 419 19770 2

**Fig. 15.1.** This illustrates the ability of a building to accommodate movement and retain integrity.

elements. This is manifested as cracking adjacent to piers in walls, which act as localized stiff zones and form points of restraint for the concentration of stresses.

Stress concentration also arises at the connections between two buildings of different rigidity and/or mass, including the junction of an existing building and a new extension. Self-weight can be an important restraining feature and lightweight buildings will be more likely to exhibit relatively unrestrained movement than heavyweight structures.

Butt ends of walls may be poorly connected. Even where they are well connected, movement in the direction normal to the span of the butting wall will not be tolerated. Either the wall will distort and crack around the junction, or a simple cleavage will form between the two, obvious by a vertical uniform crack in the internal corner.

Differential movement can also appear where damp-proof courses create slip planes within the structure. Since the DPC material transfers the forces through the structure poorly, differential movement occurs across the relatively weak plane of the damp-proof course. This can produce oversailing, which will require assessment to ensure that it is neither progressive nor unstable. Note, however, that oversailing represents a release of stresses and can therefore limit the extent of damage to the structure.

Meanwhile, highly exposed locations in the building will produce extreme conditions. This will influence the deterioration of materials and the degree of movement associated with environmental changes. The degree of restraint offered by the stiff elements of the structure will ultimately determine the response to movement forces. Hence drying shrinkage or other moisture-related movements of large-span elements, such as cast in-situ roof or floor slabs or beams, will cause deflection of the wall structure as the conversion of stresses into movement cannot be restrained. Any intermediate supports to the slab or beam will also be distorted unless there is some means of accommodating the movement. The symptoms in supporting walls running normal to the direction of shrinkage are likely to be horizontal cracks a few courses below the connection.

## Stress concentration at openings

Other stress concentrations include openings in walls. Minor openings will produce local cracking and differential movement because of the relative weakness in the structure. The transference of load around openings will mean that the walling immediately below the opening carries relatively little loading, whilst the walling separating openings carries a considerably higher loading than normal. The redistribution of these loads can produce shear cracking movement. This concentration of stresses into bands between openings where the construction is continuous occurs both horizontally and vertically. Where there is a high proportion of openings in the wall there may be a significant loss of rigidity. The wall will bulge relatively easily and offer limited resistance to racking. If there is variability in the materials within an elevation it is likely that any stresses will be relieved as cracking at the connection of materials and near the openings.

### Revision notes

- Overloading and instability are important causes of distortion in buildings.
- Distortion can manifest itself as creep or differential movement, cracking, deflection, or local detachment.
- Key determinants include:
  - the location of stresses: cracking, bulging or differential movement usually occurs at the weakest point;
  - connectivity between structures, which is especially relevant when the structures are of uneven size or mass;
  - the degree of restraint offered by the structure, which can lead to symptoms of stresses appearing remote from the initial source;
  - the location of weaknesses, including openings.

## ■ Discussion topics

- Discuss how cyclical thermal movement could manifest itself in three different constructional forms. Illustrate your answer.
- Produce a review of how openings affect the appearance of stresses in buildings.

## Further reading

Rainger, P. (1983) *Movement Control in the Fabric of Buildings*, Mitchell's Series, Batsford Academic and Institutional.

CHAPTER 16

# Wind around buildings

## Learning objectives

- You should be able to recognize those factors inherent in wind flow around buildings which can cause deficiency in the building.
- These factors have a strong structural flavour which can cause oscillation or detachment.
- You will be made aware of the complicating influence of topographical features in deficiency assessment.

## General wind effects

Wind can cause noise, generally of a high frequency when wind blows around and over projections or holes. Low-frequency noises tend to occur because of buffeting of the structure from the wind. This can cause low-frequency vibrations of the structure.

The wind can blow rain and pollutants so that they penetrate the building. They may also erode the external fabric of the building. Wind can create differential pressures over the external surface of the building, causing either positive or negative loading. This can cause wind scour. The wind scour of gravel ballast on roofs is influenced by roof slope, wind direction, size of gravel, height and the distance stones have to travel to leave the roof surface.

## Wind loading

The UK has a generally temperate climate and experiences a range of wind speeds due to the general movements of the climatic airstream and sea–land breezes. The wind is capable of sudden variations, or gusts, which can reach high speeds. These variations can cause oscillation. Framed buildings with little structural cladding or massive loadbearing walls may offer little damping to this oscillation, compared with masonry structures. Tall buildings are vulnerable since general wind speed increases with height.

The prevailing wind is from a south-westerly direction, although the geographical location will determine the intensity of wind loading. Local features around the buildings can vary considerably, making it difficult to generalize about windy environments. Hill tops are generally windy whereas built-up areas tend to slow down the wind. A failure to recognize the influence of local features may create problems.

*The Technology of Building Defects.* Dr John Hinks and Dr Geoff Cook.
Published in 1997 by E & FN Spon, 2–6 Boundary Row, London SE1 6HN, UK. ISBN 0 419 19770 2

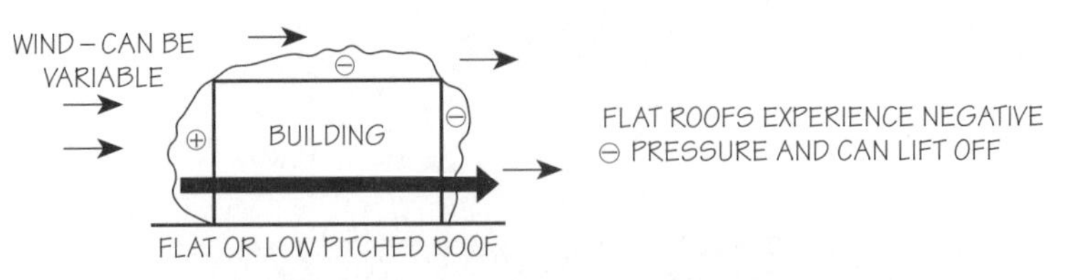


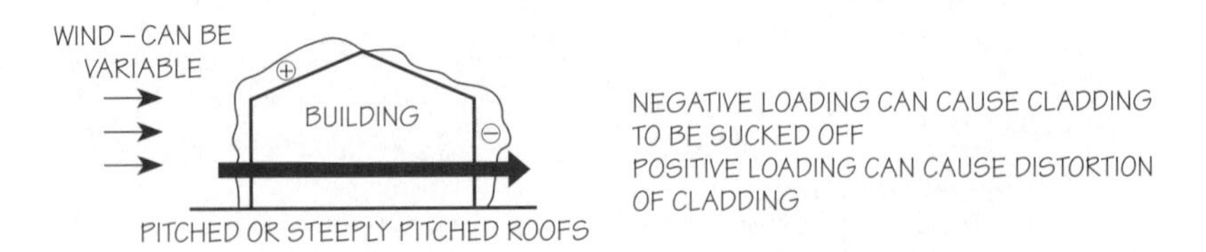


**Fig. 16.1.** General wind effects.

Since wind speed and direction have been measured over many years design guidance is only as accurate as the last storm to be included in the data.

## Orientation

The wind flow over and around buildings will create negative and positive pressure regions. Corners and changes of direction may be areas of high pressure or substantial pressure differences.

Severe positive loading can occur where the building presents a large surface area to the prevailing wind. The total wind load acting on the building due to this condition must also include any negative loading acting in the same direction. Severe negative loading may detach items from the surface of the building.

Roof tiles can move under wind action, causing the nails and/or the tiles to abrade. The tiles may blow away or drop into guttering. Ridge tiles are sited in windy locations and can be a simple measure of wind loading. Few ridge tiles have positive fixings, since they rely on their self-weight to remain in place. Poorly adhering renderings may be sucked off by the wind. Gable ends seem to be particularly vulnerable, even the outer leaves of cavity walls.

Windows may have their glass blown in or, under negative pressure effects, whole windows can be sucked out. Inadequate fixings or material failure are typical causes.

The wind loading resistance of flat roofs can be variable; they are particularly vulnerable to negative loading. Once detached they may behave aerodynamically and fly some distance. Where roofs are blown off framed buildings the remaining, now free-standing, walls may be unstable. They may then be blown over by the wind or simply collapse.

Severe loading can also occur to other features in close proximity to the building. These may include fences, garden walls and trees. Trees can be massive and are capable of inflicting considerable damage. Although trees should not be planted close to buildings, they frequently are. The shallow-rooted varieties, e.g. some of the pines, are particularly prone to being blown over.

**Fig. 16.2.** Damage caused by a tree blown over in high winds. Although the roof structure has absorbed most of the impact, there is a need to inspect for any other structural damage. (Jack Hinks.)

## Oscillation effects

A considerable number of buildings move owing to wind action. This movement is only considered critical when the structure of the buildings is threatened, or the movement becomes uncomfortable for the occupants. This appears to occur at an acceleration of between 30 and 50 /s. Tall masonry buildings have been oscillating for many years, e.g. church bell towers, and many are structurally sound. It is not unusual for horizontal and vertical cracks in buildings to open and close during the wind action.

### Revision notes

- The variable nature of wind can cause noise, blow rain and pollutants into the building and create differential pressures over the external face of the building. Local and specific features make it difficult to generalize about wind loading. Oscillation, particularly of undamped, lightweight, framed buildings can occur.
- Trees and other items can be blown onto buildings and wind can scour ballast on flat roofs. Movement or detachment of items on the external face of the building can occur. Windows can be blown in and flat roofs fly off.

## ■ Discussion topics

- Assess the influence of wind as a cause of deterioration of buildings.
- Describe a range of topographical features which may influence the effect of wind on an existing building.
- Explain why missing ridge and verge tiles are a useful guide to the wind environment around a two-storey dwelling.
- Compare the effects of the wind on defects in the external envelope of a two-storey loadbearing building with that of a 12-storey framed building.

## Further reading

BRE (1986) *Wind Scour of Gravel Ballast on Roofs*, Digest 311, Building Research Establishment, HMSO.

BRE (1990) *The Assessment of Wind Loads, Parts 1–8*, Digest 346, Building Research Establishment, HMSO.

BRE (1994) *Wind Around Tall Buildings*, Digest 390, Building Research Establishment, HMSO.

BRE (1995) *Wind Actions on Buildings and Structures*, Digest 406, Building Research Establishment, HMSO.

Buller, P.S.J. (1988) *The October Gale of 1987: Damage to Buildings and Structures in the South East of England*, Report 138, Building Research Establishment, HMSO.

Buller, P.S.J. (1993) *The Gales of January and February 1990: Damage to Buildings and Structures*, Report 248, Building Research Establishment, HMSO.

Cook, G.K. and Hinks, A.J. (1992) *Appraising Building Defects: Perspectives on Stability and Hygrothermal Performance*, Longman Scientific & Technical, London.

PSA (1989) *Defects in Buildings*, HMSO.

Richardson, B.A. (1991) *Defects and Deterioration in Buildings*, E. & F.N. Spon, London.

CHAPTER 17

# Measuring movement

## Learning objectives

You should understand:

- the nature and significance of distortion in general;
- the various techniques for measurement;
- monitoring protocols.

The visual assessment of structural movement and cracking in buildings usually focuses strongly on crack width. Cracking is symptomatic of excessive forces (generally tensile). Cracks should not be assumed to arise from a single causal defect, or indeed from a combination of defects occurring in harmony. Do not consider crack width alone, rather review the nature of the crack, the surrounding structure and the importance of the potential causes to the stability of the building. The assessment of building stability requires expert knowledge, and if you don't have it, consult an expert.

Crack width can be used in part to assist in ranking building damage according to the possible severity of the fault. It should always be borne in mind that this is a relatively simple ranking order method which allows approximate categorization of damage. They do not usually allow distinctions to be drawn between different causes of faults. Nor do they necessarily translate to the extent of dislocation in the structure or to the extent of angular distortion that may be accompanying any differential movement, both of which are important determinants of residual stability.

Assess the criticality of specific cracks according to their location in the building and their possible secondary consequences.

Cracking is caused by tensile or shear forces. The crack will appear at the weakest point or plane. It may be a plain tensile fracture, or could be compressive (for example, cracking produced on the stressed skin of brickwork panelling as it bulges under compressive forces). Cracking does not always represent the total relief of inherent stress, and frequently the cracking symptoms are physically remote from their cause.

The most direct way of assessing cracks is to monitor their behaviour. With thermal or moisture-related movements the edges of a crack may be moving frequently and cyclically. Analysis of the width may show a pattern in variations which can indicate the possible cause. Note that thermal or moisture movement of cracks is not necessarily their cause, but may simply represent (new) unrestrained movement in the damaged element.

Cracking that occurs early in the life of the building frequently involves the readjustment of the moisture contents of the various porous materials used for construction, combined with the loss of water used in the wet

*The Technology of Building Defects.* Dr John Hinks and Dr Geoff Cook.
Published in 1997 by E & FN Spon, 2–6 Boundary Row, London SE1 6HN, UK. ISBN 0 419 19770 2

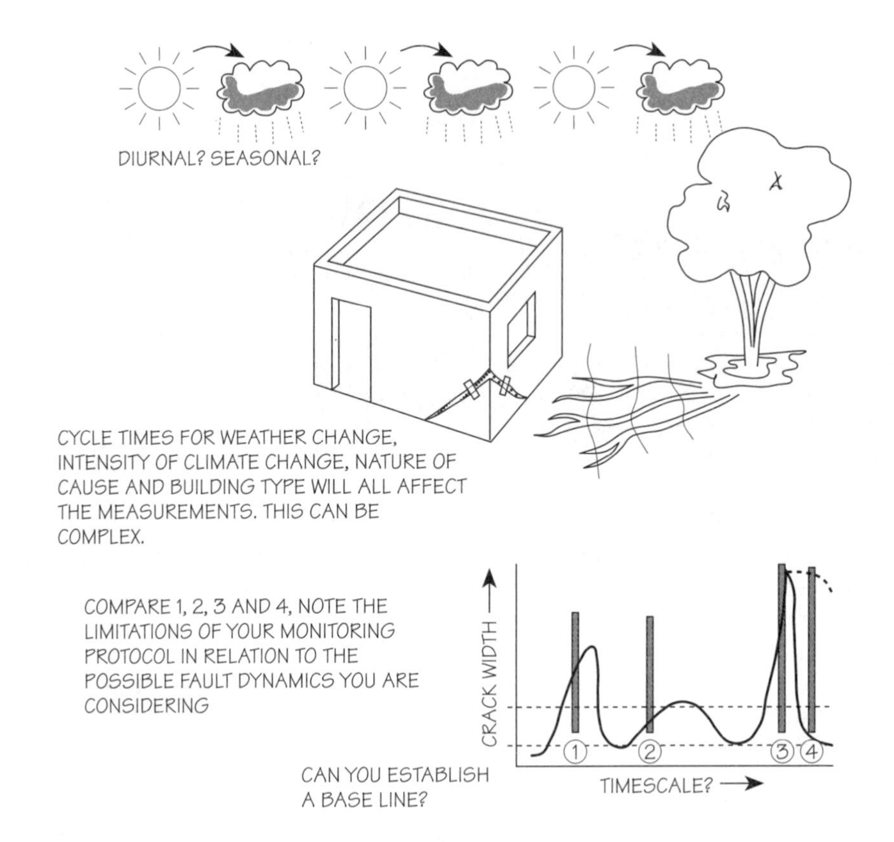


**Fig. 17.1.** Cyclic cracking patterns: the importance of measurement/ monitoring in relation to movement dynamics.

construction trades. The rate of adjustment of the building to ambient conditions will depend on the season and climatic exposure of the element, also on the patterns of use of the building. Consequently, the emergence of cracking associated with a predominance of these factors may occur in the springtime following construction during the wet season.

Progressive movement which is an immediate threat to the stability of the building is sometimes immediately obvious on first inspection.

Where the crack is stable and the cause can be identified it may then be possible to assess the residual stability of the building. There are a variety of measurement methods available, the most basic being a simple glass tell-tale slip glued onto the surfaces adjacent to the crack, which indicates movement by cracking. More modern tell-tales are graduated for temperature and allow fine measurements to be made with a reasonable degree of accuracy. They can also be used at quoins.

An alternative to tell-tales is the use of a set of markers fixed to the element and a strain gauge. Ball bearings set in epoxy resin are better than screws. These are less conspicuous and sufficiently accurate for most monitoring purposes, and can reveal rotation and shear as well as tensile or compressive movement.

Very fine measurements to a high accuracy can be made using linear variable displacement transducers (LVDTs), which convert movements into electrical signals which can be recorded automatically and at frequent intervals. The cost and potential accuracy of the system are infrequently

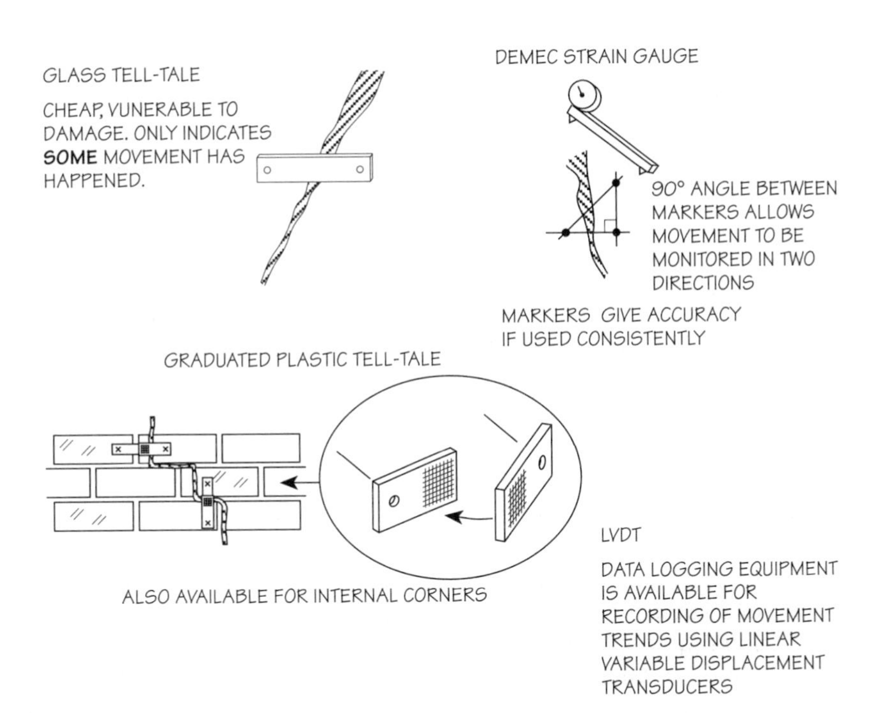


**Fig. 17.2.** Crack monitoring.

warranted. However, it does allow hands-off monitoring of frequent and infrequent movement, and could be a useful tool for monitoring buildings where an early identification of minimal movement is necessary. Combined with computerized data logging and analysis, the scope of LVDTs is impressive.

## Factors to consider when monitoring

Assessment of distortion is not merely a matter of identification and cataloguing, and in all but the most straightforward instances it will be necessary to consult a structural engineer.

Note the weather conditions at and around the time of taking measurements. If measurements are taken at the same time of day and in similar weather conditions they are unlikely to identify the trend of any thermal movement, for example.

The choice of frequency of measurement depends on the accuracy of trends required, which in turn depends on the nature of movement(s) suspected. The accuracy of the system needed will also be a factor in the choice of equipment. Ensure that the tolerance of the measurement is relatively insignificant compared with the fineness of movement expected. It may be sufficient to use the traditional glass tell-tale to monitor sample deflection. Calibrated tell-tales will give information about direction and degree of movement also. The measurements may be taken horizontally or vertically. Combining these will give information about rotation of the crack or which direction(s) the movement is in.

Unrestrained linear stresses will exhibit predominantly horizontal or vertical movement. Cyclical motion will be characterizable only after recording movements over the relevant period of time and using an appropriate number of measurements.

In cases of suspected rotation or localized sinking of the building, analyse both the top and bottom of a crack. This will allow an assessment to be made of the crack taper, which would assist in analysing hogging or sagging movements. Plain settlement cracking alone will generally minimize with time, and if a stable state is reached, it should stop moving altogether.

The criteria for acceptability of distortion, or definition of failure, must relate to the function and type of building.

Relative movement is usually more destructive than wholesale movement. These produce deformations and dislocations in the structure which redistribute loadings and can cause secondary failure in elements not designed for the new loads applied to them.

Where dissimilar buildings are connected, the dominant structure may transfer loads to the adjoining building. Low, flexible buildings attached to high, rigid structures are especially vulnerable.

A distinction requires to be made between relative vertical movement and rotational failure. Angular distortion in structures is the result of the tensile and shear components of rotational forces, and is the important criterion for the allowable deflection in loadbearing masonry. Masonry is a particularly important constructional form to assess, simply because it is a composition of numerous discrete and rigid units. The failure mode is strongly dependent on the relationship between the bricks and mortar, specifically their relative strengths and response to movement.

The relative deflection of masonry walling is expressed as a ratio of the measured deflection ($d$) to the relevant dimension of the panel, length ($L$) or height ($H$). This defines the change in shape of the section of wall as it sags or hogs.

Openings create zones of weakness in the wall. The maximum stresses that may be tolerated will depend on the location and scale of openings, the restraint of the wall and the loading it is carrying.

Where masonry is sagging, the relative deflection ($d$) is usually expressed as a ratio to the length of the panel ($L$) and shown as $d/L$. The limits of acceptable relative deformation are related to the structural connectivity to the remaining structure. The relative deflection may also be expressed using the height of the wall, for example, with hogging deflection, delta/$H$. The limits of relative deformation for hogging of walls may be half those of sagging deflections.

The relationship between the height and width of a panel will also determine the allowable relative deflection. Minimal values up to 1/300 for (sagging) acceptability in loadbearing walls and panels (Rainger, 1983) may be adjusted to include a factor of safety such as 1.5, taking the limited distortion to 1/450 (Hodgkinson, 1983). This corresponds to 11 mm over a 5 m span; however, as Hodgkinson notes, suggested limiting values are imprecise and range between 1/750 and 1/150.

It is important to distinguish between progressive and stable deformations. Distortion is important because of its effect on residual stability. Static stability is dependent on the gravitational forces of the self-weight and imposed loads of the wall across its height.

For a gravitational force the assessment of the wall is clearly discussed by a number of experts including Richardson (1988). The basic rule for design and assessment is the *t*/3 criterion. This is an expression of the effective width of the wall and the relative distortion. Essentially, if any part of the wall overhangs the wall at ground (or foundation) level by more than one-third of the thickness it is considered unstable. This may require total rebuilding of the wall, or there may be a possibility of stiffening it further.

The actual significance of overhangs between one-third and one-sixth of the thickness of the wall will depend on the degree of restraint provided by the remaining structure, and the potential restraint available for the modification of the structure. For walls to provide restraint they must be used for support in their own plane, and consequently walls parallel to a failing wall cannot be used for support. This is a common reason for the domino effect of a series of walls; tilting occurs in the bookend type of failure.

The degree of restraint posed by walls normal to the subject wall will depend on their inherent stiffness. As discussed earlier, a significant feature of this is the proportion and positioning of any openings.

Stable distortions, below about one-sixth of the thickness in overhang, may commonly be considered suitable for leaving as they are, but obviously this will depend on the specific circumstances.

Verticality and twist in structures are further features of distortion requiring assessment.

## Revision notes

- A range of fault characteristics may be used to identify indirectly the cause of defects.
- Cracks generally appear at the weakest points of a structure.
- Causes include cyclic, reversible movement and irreversible movement.
- Consider the following issues: seasonal and divisional climatic variations.
- It is essential to be able to identify the following: stable cracking, progressive movement, unrestrained and restrained movements, rotational and differential movement.
- Methods: tell-tales, strain gauge, LVDT (with data logging).
- Factors to consider when monitoring: accuracy, tolerance, relevance of weather, distortion acceptability criteria, deformation limits.

## ■ Discussion topic

- Describe how the techniques of assessing movement would differ for suggested cases of thermal movement, moisture movement, clay heave, mining subsidence and cavity wall tie failure.

## References

Hodgkinson, A. (1983) *A.J. Handbook of Building Structure*, 2nd edn, Architectural Press.
Rainger, P. (1983) *Movement Control in the Fabric of Buildings*, Mitchell's Series, Batsford Academic and Institutional.
Richardson, C. (1988) Distorted walls: survey, assessment, repair. *Architect's Journal*, **187**(2), 51–56.

## Further reading

BRE (1987) *External Masonry Walls–Assessing whether Cracks indicate Progressive movement*, Defect Action Sheet DAS 102, Building Research Establishment.
BRE (1995) *Assessment of Damage in Low-Rise Buildings with particular Reference to Progressive Foundation Movement*, Digest 251, Building Research Establishment.
BRE (1989) *Simple Measuring and Monitoring of Movement in Low-Rise Buildings. Part 1: Cracks*, Digest 343, Building Research Establishment.
ISE (1980) *Appraisal of Existing Structures*, Institution of Structural Engineers.

# Index

Page numbers appearing in **bold** refer to figures